The Future of Technical Education

John Baldwin • Neil Raven
Robin Webber-Jones
Editors

The Future of Technical Education

Ending England's Long Running Saga

palgrave
macmillan

Editors
John Baldwin
Student services
Stamford College
Stamford, UK

Neil Raven
School of Education
The University of Bristol
Bristol, UK

Robin Webber-Jones
Bedford Group
Bedford, UK

ISBN 978-3-031-70755-1   ISBN 978-3-031-70756-8 (eBook)
https://doi.org/10.1007/978-3-031-70756-8

This Palgrave Macmillan imprint is published by the registered company Springer Nature Switzerland AG.
The registered company address is: Gewerbestrasse 11, 6330 Cham, Switzerland

Contents

Notes on Contributors

John Baldwin began his career in education as a business and law lecturer in a further education college. He progressed onto being a head of department and then a director of curriculum. As part of his senior management role, he was responsible for the higher education provision in the college and its expansion. Latterly, he had responsibility for the Advice and Guidance section in the college including advice on careers and higher education. He is now semi-retired but still gives UCAS advice to the college students every autumn. With colleagues he has carried out research into widening participation in higher education and various aspects of leadership and management.

Darryll Bravenboer, PFHEA is Director of Apprenticeships and Professor of Higher Education and Skills at Middlesex University. He has extensive experience of developing vocational and work-based higher education (HE) provision in response to the needs of a wide range of employment sectors in diverse professional contexts including higher and degree apprenticeships. His research interests include the philosophy and sociology of education, HE policy, fair access, widening participation, lifelong learning, work-based and employer responsive HE.

Mandy Crawford-Lee has considerable experience of leading and negotiating skills policy, strategic planning and performance management in economic development, technical education and training reform. She is also one of the most authoritative voices on all strategic and operational aspects of higher and degree apprenticeships in England delivered by universities. Having worked with the University Vocational Awards Council (UVAC) since 2012, both inside and outside of Government, Mandy became its director of policy and operations in 2017 and, in October 2021, its first female chief executive.

Clare Dunn is the Senior Manager for Recognition of Prior Learning (RPL) at The Open University. Clare provides institutional oversight of The OU's activities in relation to RPL policy, procedures and new developments. Clare's background is in education having worked as a lecturer of sport and exercise psychology within further education, supporting students in their transition to higher education.

Neil Raven is an independent educational consultant and researcher in the field of fair access and widening participation. He is also a visiting lecturer and senior tutor in the School of Education, University of Bristol, and a visiting tutor in postgraduate researcher methods at a number of UK-based universities. Neil has edited eight books and more than 40 articles and book chapters on aspects of widening participation, evaluation planning and professional practice.

Ross Renton is Principal of ARU Peterborough, a multi-award-winning new university for Peterborough and Professor of Higher Education Policy and Leadership at Anglia Ruskin University. He has held a variety of senior roles in UK Universities and provides sector leadership through a range of external roles, including as President of Forum for Access and Continuing Education (FACE). He has a particular interest in education policy, and he is regularly invited to contribute his expertise in this area.

Robin Webber-Jones has worked in the further, higher and skills education sector for over 20 years, holding a number of senior and executive leadership roles. He has led a large number of local, regional and national projects in relation to further and higher education. He sits on a range of national sector bodies and has been a trustee of a number of educational organisations and arts-based charities. His research is focused on education leadership, educational collaborations, and social mobility and education and has contributed to a number of research publications and books. He has a keen interest in politics and has held elected roles in local government.

List of Tables

CHAPTER 1

What Is Technical Education?

Robin Webber-Jones

Preamble

Great Britain became an international power during the industrial revolution because of its superior technical expertise and training (Stephens and Roderick 1983). However, there is a popular and long-held view that technical education is inferior to academic education (Bhattacharya 2021). In this book, we examine why this view has endured and the form it now takes in offering a much needed corrective and explore why technical education retains its importance. Each chapter presents the evidence and builds the case.

Setting the Scene

Our opening preamble and Chap. 1 set the scene of the origins of technical education. It begins to define what technical education is, unpacks the issues associated with technical education and explores who is involved and why.

R. Webber-Jones (✉)
Bedford Group, Bedford, UK

J. Baldwin et al. (eds.), *The Future of Technical Education*,
https://doi.org/10.1007/978-3-031-70756-8_1

The Challenges

Chapter 2 explores the complicated way in which technical education has emerged through a wide range of policy developments. It sets out the history of policy making, and how this coalesced in more recent times around apprenticeships and T Levels. It then draws from research with practitioners to investigate the impact and development of T Levels in the further education sector.

We explain that although T levels are described as being technical qualifications some are not and all of them are more academic than the qualifications they are replacing. Consequently, their academic content may dissuade some of those they are aimed at from taking them, including, crucially, young people from more deprived backgrounds not only due to the nature of the provision but also the costs of accessing the compulsory work experience component.

Challenges and Responses

Against the backdrop of concerns about low productivity growth and rising skills shortages across England, Chap. 3 argues that Further Education colleges (FECs) should be key players in delivering higher technical qualifications, and that a collaborative approach to curriculum development offers a more effective method of doing this than one based on competition between local FECs. This case is made through exploring the outcomes of a regionally based collaborative project involving nine FE colleges as well as employers' organisations and local authorities. However, to sustain and build on the success of this initiative there is a need it is argued, for adequate long-term funding and a period of policy continuity.

The main theme in Chap. 4 is concerned with the lack of attention paid to opportunity for apprentices to progress on to higher education after completing their study. It recognises the difficulties in applying credit and UCAS points to level 3 apprenticeships because of their varied content and time to complete but provides a method of ascribing credit and thus removes a barrier to apprentices taking HE study.

Chapter 5 examines the charge that degree apprenticeships are not contributing to social mobility because they are being taken up mainly by the middle classes. The author refutes this allegation and gives examples of how they are being used to advance social mobility by police forces who are increasing the number of apprentices from underrepresented ethnic backgrounds, by encouraging more of those over 25 to take HE study and she provides statistics to show that at some universities the proportion of young people from deprived areas on engineering and digital degree apprenticeships is increasing.

The themes that emerged from Chap. 6 which relate to social mobility and technical education include that a broad range of underrepresented groups are generally less likely to access further or higher technical or academic education. Some groups such as those from deprived backgrounds might be thought to take up technical apprenticeships but the number doing so is declining and those form better-off backgrounds are in some technical professions taking up apprenticeships. Many disabled people are less likely to take up technical education courses but others such as those with autism are more likely to. We explain that with the advent of the Lifelong Loan Entitlement, more adults should be able to access technical education to upskill.

Chapter 7 unpacks the themes that have emerged earlier in the book and suggests how technical education will evolve in the next 30 years using information gleaned from employers and those working in the education sector. In presenting a potential future scenario it draws upon six key themes that are integral to a long-term economic plan but also indicates the barriers to successfully realising a preferred policy landscape. It then presents the key policy levers that will be needed to create an integrated policy framework for tertiary education which ensures technical education fits seamlessly into the wider system.

A key theme running though this book is that whilst policy makers say they understand the importance of technical education, the policies and practices they put in place demonstrate a limited real understanding of its role and potential in sustaining and boosting the economy. Instead, technical education is viewed as just another form of academic education without recognising the importance of craft skills as well as knowledge. They rightly see the role of employers as being critically important but do not harness their knowledge and expertise effectively and only involve a limited number of employers' at a cursory level—partly because many employers do not see the provision of education and training as important. In 2022, 40% of employers did not provide any training for their employees (Gov UK 2022).

Part I Defining Technical Education

Imagine two individuals sat at the entrance to the British Library, drinking coffee and debating what technical education is. The discussion tries to put clear parameters around what it is. They imagine the books and archives contained within the building helping them. Would a degree in medicine be considered as something technical? After all it is practical, involves problem solving, there are distinct pedagogies that can be used to teach it, there is a clear career structure linked to developing good levels of competency. Yet to access it they both smile at the fact people need 'academic' qualifications. The perception is that it is one of the most challenging academic routes and yet those that become doctors undertake their work often as a vocational calling, undertaking a whole series of practical activities. They reflect on how it is considered one of the oldest professions. They recall the research they read on whether or not law should be included as part of the University curriculum and how this linked to providing qualifications that could be taken forward by a profession. There is already a sense between them that 'technical' education may not be a new thing and that there are challenges regarding the precision of what it is.

The discussion continues to focus on how the study of 'technical' things is different to other things. They conclude that it is 'technical' unless it is an 'A' level or equivalent at level 3. They then continue to discuss this and conclude that even this is not a satisfactory definition—where would A Level physics for example fit—surely there are technical elements and the link between theory and practice means it can applied. They consider the ways in which higher education is currently 'measured' by elements such as graduate employability, and how the newspapers consider the 'worth' of higher education. They discuss specific types of educational programme and focus in on apprenticeships. They contemplate the different 'levels' that they are available on and discussions in parliament and the press regarding the most appropriate level that an apprenticeship should be offered at. While there has been a steady uptake in degree apprenticeships since the introduction of the apprenticeship levy, while those at levels 2 and 3 have seen a decline.

In the end, they give up and purchase another cup of coffee. The only thing that they can conclude is that it is complicated and that the discourses used to attempt to define it are not as clear as they are often made out to be.

Definitions and Esteem

This discussion in many ways shines a light on a number of issues which start to articulate the complex technical education landscape. It is a landscape which research concludes has several characteristics which can often seem to be:

- There has often been a struggle for technical education to achieve a parity of esteem with academic education. 'Academic' often means 'A Levels', the subjects that make up the 'Ebacc' in the school's curriculum and traditional University degrees.
- Yet increasing the volume of technical education has been linked extensively to increasing productivity especially following the 2008 financial crash. This is partly because other countries with more mature higher-level skills systems grew more rapidly following the financial crash. There is some context around their approach to austerity and government investment that is not always contextualised in relation to this. Prior to this period, the term 'vocational' was often used to describe this type of education. However, this had further connotations to it which many in the sector considered to be negative.
- In this there has been research conducted by Unwin (2006) and others on how the international technical education landscape provides more scope for progression, lifelong learning, and the accumulation of higher-level skills in other countries. Willets (2017) and others attribute this to the earlier emergence of 'technical' universities in Europe that universities linked to trade guilds in England.
- There is a sophisticated regulatory landscape which people working in technical education find very complicated, and therefore costly.

Therefore, it is important to set out how we might start to define technical education, how technical education as we understand it today in England has emerged, and how effective it is in achieving its stated ends. The context of this is important. Hyland, reminds us that there has often been a perceived lack of parity of esteem between technical education and other academic routes:

> The perennial problems of vocational education....consisting in the subordinate and inferior status of vocational studies as against academic pursuits – has led to a range of difficulties and anomalies which have bedevilled progress in the field for a century or more. (Hyland 2017, 306)

The History of Technical Education

The challenge is that the term 'academic' is equally not understood as something concrete. However, the history of technical education in England is characterised by a series of developments and changes that reflect the evolving needs of society and the economy. Here's a brief overview of key milestones:

1. Nineteenth Century: The Industrial Revolution
   - The early nineteenth century saw the beginning of the Industrial Revolution in England, leading to significant technological advancements.
   - The need for skilled workers in industries such as manufacturing and engineering became apparent.
2. 1837: The Factory Act
   - The Factory Act of 1833 included provisions for the education of factory workers' children. However, these early efforts were more focused on basic education than technical skills.
3. 1861: The City and Guilds of London Institute
   - The City and Guilds of London Institute was established to improve the training of craftsmen, technicians, and engineers.
   - It played a crucial role in the development of technical education and the certification of skilled workers.
4. 1870: Forster Act
   - The Elementary Education (Forster Act) 1870, marked a significant step in the development of general education in England. It established the framework for elementary education but did not specifically address technical education.

5. 1889: The Technical Instruction Act
    - The Technical Instruction Act, 1889 also known as the Mundella Act, allowed local authorities to levy a tax to fund technical and manual instruction.
    - This act aimed to promote technical education and bridge the gap between academic and vocational training.
6. 1900s: Growth of Technical Colleges
    - The early twentieth century saw the establishment and growth of technical colleges, which offered a range of courses in technical and vocational subjects.
7. 1944: The Education Act
    - The Education Act of 1944, also known as the Butler Act, introduced the tripartite system, dividing education into three stages: grammar schools, secondary modern schools, and technical schools.
    - However, the implementation of technical schools was limited, and many areas did not establish them.
8. 1960s–1970s: Rise of Comprehensive Schools
    - The comprehensive school system, which aimed to provide a more equal and inclusive education, led to a decline in the number of technical schools.
9. 1980s–1990s: Shift towards Vocational Education
    - Economic changes and a growing emphasis on vocational education led to the establishment of vocational qualifications and the promotion of apprenticeships.
10. Twenty-first Century: Further Reforms
    - The twenty-first century has seen ongoing efforts to reform technical education, with a focus on aligning skills training with the needs of the modern economy.
    - Initiatives such as the introduction of apprenticeship standards and reforms to technical education qualifications have aimed to improve the quality and relevance of technical training.

11. 2017: Introduction of T Levels
    - T Levels were introduced in England as a technical alternative to A Levels, providing students with a more focused and practical route to employment or further education.

As can be seen, the history of technical education in England reflects a dynamic interplay between societal needs, economic demands, and educational policies.

Apparent contrasts between academic and vocational education or between higher education and further education can also generate confusion and conflict, but actually they are often closely linked and complementary. David Hume put the point beautifully:

> An advantage of industry and of refinements in the mechanical arts is that they commonly produce some refinements in the liberal; nor can one be carried to perfection without being accompanied, in some degree, with the other. The same age, which produces great philosophers and politicians, renowned generals and poets, usually abounds with skilful weavers and ship-carpenters. We cannot reasonably expect that a piece of woollen cloth will be wrought to perfection in a nation which is ignorant of astronomy or where ethics are neglected. (Hume 1752)

These policies start to fill the conversation that take place between our two colleagues at the library. Conversation moved swiftly to what is considered as the purpose of education; whether its outcomes are (or should be) simply linked to economic activity or whether there are broader purposes for education. They start to question:

- How are teaching and learning approaches developed?
- How employers and other engage in deciding what is worthwhile.
- Whether or not there is anything relating to assessment practices that influences technical education
- What the difference between 'technical' and 'vocational' is.

Both are interested in the history of education policy—this might explain the setting they are in. They start to debate the origin of the link between skills and education as it is framed today. While it is evident that the mediaeval guilds had a role to play in skills education, it was not until Prime Minister Callaghan's 1976 speech at Ruskin College Oxford where purpose of education was more closely linked to economic output (Callaghan 1976). This followed the expansion of 'vocational' universities which emerged from the 1900s. They had emerged earlier in Europe but the system in England prevented earlier expansion.

Recent Changes in Technical Education

After spending time working across the further and higher sectors, our two reflective friends start to consider the technical education landscape in more recent times. After another cup of coffee, they conclude that the vocational education landscape continues to expand under successive government free market approach to education. However, the coalition government in 2010 takes (what they would argue as ideological) decisions to cut the size of the state. This led to:

- **Tuition Fee Increase:** One of the most controversial decisions was to raise the cap on tuition fees for undergraduate courses in England. The cap was increased from £3000 to £9000 per year, leading to widespread protests and debates. This policy aimed to shift a larger share of the cost of higher education from the government to students.
- **Higher Education Funding Changes:** The government implemented changes to the way higher education institutions were funded. Funding for teaching in universities became more reliant on tuition fees rather than direct government grants.
- **Student Loan System Changes:** The coalition government introduced changes to the student loan system. The threshold for repaying student loans was increased, and the repayment period was extended. This was seen as an attempt to make higher education more accessible by reducing the immediate financial burden on graduates.

- **Maintenance Grants and Loans:** There were changes to the system of maintenance grants and loans for students from lower-income families. Maintenance grants, which were non-repayable, were replaced by larger maintenance loans, which students were expected to repay after graduation.
- **Further Education Funding:** The coalition government also made changes to funding in further education, including reductions in funding for adult education courses. There were efforts to redirect resources towards courses that were deemed to have greater economic relevance and were more likely to lead to employment. (Department for Business, Education and Skills 2010)

These funding changes were part of broader efforts to address the economic challenges facing the country and reduce public spending. However, they were met with significant criticism, particularly from students and educators who argued that the increased financial burden on students could limit access to higher education for those from less affluent backgrounds.

Our friends in the library have worked closely within the sector and they have been exposed to a range of 'grey literature' in relation to the sector. They reflect on the impact of higher fees in HE and the 'Dual Mandate' where further education colleges have their defining purpose as 'community engagement' and higher level technical and professional courses. In many ways, they consider this to be the ongoing framework within which technical education operates. Through their conversation they note that:

- The term 'technical and professional' emerged with the work AOC undertook, with Gatsby, to think about how to position 'vocational' education, develop alterative higher level learning routes (in response to the ESS Workforce 2022 Report) (DfE 2022).
- During this time, the TVET (technical and vocational education and training) approach had been developed by UNESCO (and continues to this day) so the term 'technical' provides some international recognition which Gove and others were keen to be seen to do from 2011 (see 'the Importance of Teaching', for example).

- The one phrase which has been constant is 'apprenticeships' but their form and structure have altered significantly through the levy and through a move towards higher-level skills being a focus of the apprenticeship landscape. This has been transformative and internationally leading, but the name has stood as something understood.
- The models of employer contribution and higher level continue to influence the 'technical and vocational' policy landscape with new local skills improvement plans (LSIPs) exploring employer contribution to technical education. It looks as though technical education will continue to be the subject of short-termism as approaches like LSIP, Univoc and so on.

They consider if the word 'technical' can be defined through the eyes of the United Nations who say: 'TVET comprises education, training and skills development relating to a wide range of occupational fields, production services and livelihoods. It empowers individuals, organizations, enterprises and communities and fosters employment, decent work and lifelong learning thereby promoting inclusive and sustainable economic growth and competitiveness, social equity and environmental sustainability' (UNESCO 2023).

Or they consider if we can view it by what it is not. For example, the Tomlinson Report 2004 tried to create a Baccalaureate type system for England, but it was rejected because of the impact it would have on A Levels. The political fallout would have been too great. So Technical education in England could be anything that is not an A Level—although as they had already discussed—even this divide does not work well.

Judgement on the quality of teaching and learning also become part of the discourses that affect technical education, and they spend time considering this. They consider that:

- The opening up of discussion on pedagogies where summative assessment is not exam based, means technical 'approaches' to teaching and learning are growing—as demonstrated by an emphasis on employer co-creation, problem-based learning, project-based learning, real-world learning, experiential learning and so on. They consider the wide range of organisations who are involved in research in these approaches currently.
- On this basis apprenticeships, HTQs, T Levels, BTECs (and similar) all form part of the structures in the complicated 'technical education' landscape. Qualifications and programmes of study in this way

also have multiple sites where they can be taught (colleges, workplaces, community centres, etc.), which is also a point of contrast to 'academic' approaches.

Essentially, our friends conclude that the landscape for technical education in England is complicated. They consider how much more complicated it is becoming through a range of reforms including T Levels, changes to apprenticeships and, at the time of the conversation the proposed introduction of the 'Advanced British Standard' (DfE 2024). While these may start to develop the discussion and dialogue about what technical education is they add to the complexity.

Attempts have been made to distinguish the term, 'vocational education' from 'technical education'. Cantor (1985) distinguishes between vocational 'education' which is provided in schools, colleges and universities as part of the educational system, and vocational 'training', which is preparation for work provided for and in industry. However, today schools, colleges and universities would suggest that they deliver vocational education but as we see later colleges deliver T levels which are defined as technical qualifications. Gleeson and Mardle (1980) suggested that 'technical' education is usually taken to refer to the mastery of a body of knowledge and skills, which can be applied in a practical way. In much of the literature the two terms are often used interchangeably.

The consistent changing and challenge of the political positioning of technical education is one thing. Understanding the current structures is another. The first thing to note is that, since the financial crash of 2008, there has been a significant challenge in the British economy of low productivity. There has also been, perhaps understandable, pressure from the Treasury to reduce the 'churn' in individuals repeating the same level of qualification. Therefore, for this reason (and other reasons which will be unpacked further in the coming chapters) most of the reforms to advance technical education have taken place. Many of these reforms have focused on work undertaken by a body which is now known as the Institute for Apprenticeships and Technical Education (IFATE). They have employer bodies who articulate 'standards' that articulate the 'knowledge', 'skills' and 'behaviours' that employers want to see against a range of priority job roles. These standards are then used as the basis for all apprenticeships for T levels (Skills and Post 16 Education Act 2022). At Levels 4 and 5, the equivalent of a 'T' level is a 'higher technical qualification'. As can be imagined, there is significant machinery of state placed around these

Table 1.1 Technical education bodies involved

Element	*Description*
Education and Skills Funding Agency (ESFA)	Part of the Department for Education. It oversees technical education policy implementation and they oversee key parts of skills policy such as the sub-contracting standard, and general performance. A monthly Individual Learner Record (ILR) is sent to them to detail current student numbers, performance and income.
Institute for Apprenticeships and Technical Education (IFATE)	The Institute has a number of roles in relation to technical education including the establishment of the aforementioned standards. It also has a role in overseeing endpoint assessment practices for apprenticeships.
Office for Students (OfS)	The Office for Students Regulates Higher Education. Higher and Degree Apprenticeships are included in elements of its regulatory framework and it does have some responsibility to oversee integrated endpoint assessment where they apply in degree apprenticeships.
Ofqual	This is the qualification regulator. It approves standards and assessment regulations for qualifications predominantly on qualification frameworks—NQF, RQF and QCF. It has a role with awarding organisations to ensure that their qualifications and assessment practices are sound. For T Levels it awards the role of which awarding organisation should run the qualifications for each T Level.
Office for Standards in Education (OfSted)	Ofsted inspects all skills and apprenticeship providers using its framework.
Apprenticeship Levy	The levy is paid by employers who have a payroll of more than £3 million currently. The levy (at the time of writing) can then be used to pay for apprenticeships. There are some elements of levy that can be transferred to other partners if an employer is not going to spend all its levy in the given timescale.

things including a qualification and awarding structure, assessment methodologies, thematic inspections, summative assessment, and there are roles that regulators and funding agencies play.

For those who lead and operate across the technical education landscape they find complexity. Even the way it is funded and operates is not simple. Examples of the bodies involved are included in Table 1.1.

While many already working in the technical education will recognise the complexity shown in this table and will be involved in trying to navigate all the bureaucracy that goes with it.

Part II Teaching and Assessing Technical Education

How does the teaching and assessing of technical education differ from the teaching and assessing of academic education? Technical education is delivered at all levels of education from foundation level to master's and PhD level. It is delivered at all ages from primary school though to university. It is delivered by a wide range of providers, not only in schools, colleges and universities but also in prisons, by private specialist providers, charities and most employers. The qualifications in technical education are provided by a very wide range of qualification bodies, some very large and some very small. It is delivered on a full-time, part-time and distance learning basis and classroom or workshop based or a combination of the two such as apprenticeships. Consequently, identifying how technical education is taught and assessed is difficult because of this wide range of variables but also the complexity and depth of technical /vocational areas differs markedly. We would suggest that academic education is not so complex and wide ranging.

Teaching

However, there are some basic principles that apply to the teaching and assessing of technical subjects. Lucas et al. (2012), indicate that there is very little written or discussed about vocational pedagogy because there is lack of clarity about what technical education is for. Much of it is delivered in Further Education Colleges (FECs) and is continually changing and subject to managerial influences and thus is seen as meeting the skills requirements of business and industry. Moreover, the teachers have a dual professional identity as they are expected to be skilled in the area they are delivering but also be qualified teachers and sometimes these two identities clash. In addition, there are inadequate models of vocational education and the methods used to teach tend to be narrow and limited and based on general teaching methods rather than being adapted for vocational subjects. They also suggested that there are poor analogies for vocational education. It is sometimes seen as being 'sitting with Nellie' and newer methods such as problem-based learning may be used as a bolt-on rather than being integrated into vocational teaching. Finally, Lucas, Spencer and Claxton found that vocational teachers tend to be reluctant to use theory—either because there is little theory in vocational education to refer to or because vocational teachers tend to be more pragmatic and used to learning by their own experience.

Young (2004) argues that the wide range of technical subject areas makes if difficult to decide how they should be taught. He suggests that the following need to be taken into account when doing so.

- The balance between subject knowledge content and workplace practical knowledge.
- The extent to which the subject is established and has an agreed body of content or workplace processes. Some technical areas have been established for centuries and some are only just emerging.
- Should a competence or outcomes method of delivery be used?
- The value of generic pedagogic skills that can be used across a range of technical areas and how easily they can be integrated into the subject being taught.
- The balance between general pedagogy and vocational pedagogy where general areas of knowledge such as maths need to be integrated into the teaching.

In order to attempt to distinguish between different types of technical education and how it should be taught, Lucas et al. (2012) suggest three categories—Physical Materials, People or Symbols. Physical Materials cover subjects such as hairdressing, plumbing, bricklaying and make up. People covers areas such as nursing, retail, hospitality and care, while Symbols covers graphic design, accountancy, journalism and games design. They suggest that the pedagogy relating to each category may have something relative to each of the crafts or professions but that those teaching in Symbols areas for instance, should not forget about the People and Physical Materials relevant to their area. However, they go on to argue that every technical area has six outcomes which need to be taught or developed:

- The routine expertise required for that technical area—being skilful
- Being resourceful
- Having functional literacies such communication, numeracy and IT skills
- Craftsmanship—having pride in a job done
- A business-like attitude to work—commercial sense
- Wider skills for growth and lifelong learning

They then suggest the types of teaching methods which are best adapted for teaching technical education although most of these can apply to teaching academic education too. These are:

Learning by watching—This is often used when teaching crafts such as catering, plumbing or hairdressing where the tutor gives a demonstration first and then challenges the students to carry out the task.

Learning by imitating—Having watched the teacher or another professional the student then attempts to replicate what they have seen, often used in nurse training.

Learning by practicing—Once a student has grasped a skill, this is the frequent repeating which leads to mastery and would include craft subjects such as carving but also skills relating to sport, sales and teaching.

Learning through feedback—This can be written or spoken and if provided skilfully and sensitively will lead to a student becoming more knowledgeable and skilled.

Learning through conversation—Putting novices with experienced practitioners and the relating of the best methods of working. This has been used with experienced and trainee midwives.

Learning by teaching and helping—This relates to peer learning where students who are more able on one learning task can help others but be helped themselves on other tasks.

Learning by real-world problem solving—This applies to the mastery of skills by having to solve real problems. It was initially recognised in medical training but Farrar and Trorey (2008) recognised it with dry stonewallers who built up innate skills over time.

Learning by critical thinking—This is a skill that employers want their employees to have. Although it is difficult to teach, it can be developed by frequent challenging of students to think deeply about problems and teaching them how to do that.

Learning by listening, transcribing and remembering—Although challenging students to learn by experience is effective, it is recognised that in all technical areas there is a bank of basic knowledge and skills that are needed, and which should be taught by listening to a skilled teacher (Kirschner et al. 2006).

Learning by drafting and sketching—This is a skill used in teaching architecture and art subjects but can also be used in the form of rough diagrams or doodles to help thinking and understanding by learners in most subjects.

Learning by reflecting—This is a form of experiential learning where students are encouraged to reflect on skills they have gained and tasks they have completed to repeatedly examine if they can be improved (Kolb 1984).

Learning by being coached—This is a prime learning method in sport but is used increasingly in management training, particularly in areas such as construction where there is an aging senior workforce who can pass their knowledge and experience onto younger colleagues.

Learning by competing—Constructive competition can help a learner to refine and improve their skills. This can be within organisations or at a national or international level. In the UK Worldskills final of 2023, the finalists competed in 51 skills, ranging from digital construction, health and social care and additive manufacturing to metal fabrication and cyber security.

Learning through virtual environments—This exciting emerging method of learning is being used in areas such as medicine and engineering where students can practice in safety in an authentic environment such as an aeroplane engine but in the classroom or at home.

Learning by simulation—This can be used in nursing or in the police or fire service to test out and enhance teams' abilities to deal with emergencies in a simulated environment.

Learning through playing games—This can be a valuable method of team building and enhancing learning in communication and working with others.

Technical education has to take place in a suitable learning environment as it can be classroom based but is often in a simulated workplace environment such as a kitchen, workshop or beauty salon, which can often mean that technical education has more expensive capital costs and thus is often based in FE colleges or universities rather than schools.

Vocational learners can often be very motivated, particularly where they have become disenfranchised from learning at school through lack of interest but to harness that interest the teacher has to be very skilled. As mentioned earlier, those who teach technical education have a dual professional identity. They need to be knowledgeable and skilled in the subject they teach but also have to have teaching knowledge and skills. There can also be dissonance between the culture, mores and assumptions of the professional background and the mores and assumptions of a teaching identity. Page (2013) examined the difficulties that construction lecturers have in FE colleges where previously working in the male-dominated industry, they were expected to be tough and often politically incorrect and when they became lecturer they were expected to be caring and very careful in the language they use in front of students. The writer as head of a hospitality and catering department in a FE college had to remonstrate with the chef lecturers who were used to regularly swearing in their previous roles and who justified their language at college by arguing that the students needed to be aware of the environments they were going to be going into when they finished their course. They found it difficult to break the habit. Moreover, Cowen and Wilson (2022) found that in FE colleges and work-based learning providers teaching technical education, there can be up to eight roles involved in the delivery of learning, teacher, learning coach, instructor, technicians, programme leaders, course leaders, expert teacher and progress tutor, far more than are required in the delivery of academic education. This variety of roles is partly because of the need for teachers of technical education to often work closely with employers, partly because it is their employees they are teaching and assessing but also to ensure the teachers keep up to date with current occupational knowledge and skills.

As mentioned previously, much technical education takes place in FE colleges at levels 1 to 5, but colleges find it very difficult to recruit lecturers with both the technical and teaching knowledge and skills. Bates (2022) suggests that there are a number of reasons for this. The pay of FE lecturers is lower than that for equivalent schoolteachers and HE lecturers and thus those who come into teaching from business or industry often take a pay cut. HE argues that this is because in real terms the spending on adult education and apprenticeships in FE is about 50% lower than it was in 2009/2010. Bates suggests that the reforms being discussed in this book will not be successful unless more investment is put into technical education.

Assessing

How does the assessing of technical education differ from that of academic education? Most of academic education has traditionally been based on examinations from SATS through to master's level. However, assessment in technical education is different, although exams are often used. Field (2021) suggests that although the assessment of technical education is critical, knowledge and research of it is very weak. This is partly because of the diversity of the occupations involved in technical education and the wide variety of assessment methods that are used. He suggests that although the evidence gleaned from a knowledge-based exam may be what university entrance requires, this may not be what an employer needs to assess whether a person is competent to carry out the tasks needed in the messy realties of working life. This is also slightly complicated by the fact that many young people use vocational qualifications such as BTEC Extended diplomas to progress to university.

Field argues that there are a number of factors which must be taken into account when assessing technical education:

Balancing authentic work tasks with standardisation—An employer wants to know whether the prospective employee can carry out the tasks needed in that particular workplace but at the same time there needs to be some standardisation of the assessment. Therefore, assessment needs to be a combination of the assessment of authentic work tasks with written or oral examinations.

Using synoptic assessment—Using realistic projects to not only assess the specific workplace knowledge or skilled needed but also to assess problem-solving skills and the ability to deal with challenges.

Using periodic assessment—Rather than only depending on endpoint assessments, periodically assess using a range of assessment methods.

Implementing grading—When assessing specific competencies, it is often usual to only give a pass or no grade because a student can do the task, or they cannot.
However, most tasks can be carried out to various levels of competence although it is more difficult than marking a maths exam when an answer is correct or not.

Quality assured assessment—An employer may not only need an assurance that a student has passed a series of assessments but that they have also fully completed a training programme or have undertaken a period of work placement.

Improving the evidence base—Field suggests that there is very little research on which type, method and timing of assessment provides a robust reliable outcome. Newton (2018) reviewed the journal *Assessment in Education* and found that out of 400 articles published over the period of a decade, just two related to technical and vocation assessment.

This section on teaching and assessing technical education has sought to provide a limited introduction which will be broadened in subsequent chapters.

Conclusion

In this chapter we have discussed the history of technical education, its vital importance today, how it is defined and distinguished from academic education and how the teaching and assessing of technical education differs from that in academic education.

References

Bates, N. 2022. Without Proper Investment in the Teaching Workforce, England's Technical Education Reforms Are Doomed to Failure. *FE News*. https://www.fenews.co.uk/exclusive/without-proper-investment-in-the-teaching-workforce-englands-technical-education-reforms-are-doomed-to-failure (Accessed 14 January 2024).

Bhattacharya, A. 2021. *Not Just Other People's Children: What the Public Thinks about Vocational Education*. Social Market Foundation. Available at https://www.smf.co.uk/publications/public-attitudes-to-vocational-education/ (Accessed 4 June 2024).

Callaghan, J. 1976. *A Rational Debate Based on the Facts*. Available at https://education-uk.org/documents/speeches/1976ruskin.html (Accessed 8 April 2024).

Cantor, L. 1985. Vocational Education and Training: The Japanese Approach. *Comparative Education* 21: 67–75.

Cowen, G., and P. Wilson. 2022. *Research into Technical Education Teaching Roles*. London: Gatsby Foundation.

Department for Business, Education and Skills. 2010. *Changes to Tuition Fees and Higher Education*. Available at https://www.gov.uk/government/news/changes-to-tuition-fees-and-higher-education (Accessed 8 April 2024).

Department for Education. 2022. Employer Skills Survey. Available at https://explore-education-statistics.service.gov.uk/find-statistics/employer-skills-survey/2022 (Accessed 9 April 2024).

———. 2024. *First Step Towards Introducing the Advanced British Standard*. DfE.

Farrar, N., and G. Trorey. 2008. Maxims, Tacit Knowledge and Learning: Developing Expertise in Dry Stone Walling. *Journal of Vocational Education* 60 (1): 35–68.

Field, S. 2021. *A World without Maps—Assessment in Technical Education*. London: Gatsby Foundation.

Gleeson, D., and G. Mardle. 1980. *Further Education or Training? A Case Study in the Theory and Practice of Day Release Education*. London: Routledge and Kegan Paul.

Gov UK. 2022. Employer Skills Survey. Available at https://explore-education-statistics.service.gov.uk/find-statistics/employer-skills-survey/2022 (Accessed 5 June 2024).

Hume, D. 1752. *Political Discourses—Of Refinement in the Arts*. Available at https://davidhume.org/texts/pld/ra (Accessed 8 April 2024).

Hyland, T. 2017. Craft Working and the "Hard Problem" of Vocational Education and Training. Scientific Research Publishing.

Kirschner, P., J. Sweller, and R. Clark. 2006. Why Minimal Guidance During Instruction Does Not Work: An Analysis of the Failure of Constructivist, Discovery, Problem-based, Experiential, and Inquiry-based Teaching. *Educational Psychologist* 41 (2): 75–86.

Kolb, D. 1984. *Experiential Learning: Experience as the Source of Learning and Development*. Englewood Cliffs, NJ: Prentice-Hall.

Lucas, B., E. Spencer, and G. Claxton. 2012. *How to Teach Vocational Education: A Theory of Vocation Pedagogy*. City and Guilds of London Institute.

Newton, P. 2018. Grading Vocational & Technical Qualifications: Recent Policies and Current Practices, Ofqual. In https://assets.publishing.service.gov.uk/government/uploads/system/uploads/attachment_data/file/755681/Grading/vocational/and/technical/qualifications (Accessed 14 January 2024).

Page, D. 2013. Recruitment and Transition of Construction Lecturers in Further Education: The Perspective of Middle Managers. *Educational Management Administration & Leadership* 41 (6): 819–836.

Skills and Post 16 Education Act. 2022. Available in https://bills.parliament.uk/bills/2868 (Accessed 9 June 2024).

Stephens, M.D., and W.R. Roderick. 1983. The Role of Scientific and Technical Education in Early Industrialisation in America and England. *The Vocational Aspect of Education* 35: 90.

The Elementary Education Act. 1870. Available in https://www.parliament.uk/about/living-heritage/transformingsociety/livinglearning/school/overview/1870educationact/ (Accessed 8 April 2024).

The Factory Act. 1833. Available in https://www.parliament.uk/about/living-heritage/transformingsociety/livinglearning/19thcentury/overview/factory-act/ (Accessed 8 April 2024).

The Technical Instruction Act. 1889. Available in https://education-uk.org/documents/acts/1889-technical-instruction-act.html (Accessed 10 April 2024).

Tomlinson, M. 2004. *14–19 Curriculum and Qualifications Report*. DfES. Available at https://education-uk.org/documents/pdfs/2004-tomlinson-report.pdf (Accessed 9 June 2024).

UNESCO. 2023. *Global Inventory of Regional and National Qualifications Frameworks*. UNESCO. Available at https://unesdoc.unesco.org/ark:/48223/pf0000386929 (Accessed 9 April 2024).

Unwin, L. 2006. *Vocational Education in Everyday Life*. London: Routledge Falmer.

Willets, D. 2017. *A University Education*. Oxford: Oxford University Press.

Young, M. 2004. *The Importance of Vocational Pedagogy*. Presentation to the Research Seminar: Vocational Pedagogy, 22 September 2004. City & Guilds.

CHAPTER 2

Practice Informed Policy

Robin Webber-Jones

Introduction

This chapter will set out the complicated way in which technical education has emerged through a wide range of policy developments. It will set out the history of policy making, and how this coalesced in recent times around apprenticeships and T Levels. It then draws on research with practitioners on the impact and development of T Levels in the further education sector.

The Development of Technical Education Policy

In Chap. 1, the history of technical education in England was discussed, ranging back to the mediaeval guilds from about 1250, where part of the role of the master was to educate and train the apprentice in the craft (Evans 2022). In each century, the need for technical education has continued, including in 1571 when a scheme was established in Norwich to teach techniques in the manufacture of clothing, in 1663 when the Royal Society was established to improve natural knowledge and in 1760 the Mechanics Institutes movement began (Evans 2022a). However, it was

R. Webber-Jones (✉)
Bedford Group, Bedford, UK

J. Baldwin et al. (eds.), *The Future of Technical Education*,
https://doi.org/10.1007/978-3-031-70756-8_2

only with the Technical Instruction Act 1889 that the government realised that to compete with other European countries state involvement was needed (Green 1995). Under this Act, local committees were formed to create institutions that could provide technical education. The Education Act 1902 created Local Education Authorities (LEAs) who administered schooling including technical education, but the law was permissive and thus led to a wide variation in provision. The Education Act 1944 created a formal further education sector created by the Education Act 1944, which tasked LEAs with providing adequate full- and part-time provision for post school-age students and a shift began from evening study into full-time study and the creation of more technical colleges. However, the provision differed markedly depending upon each LEA's priorities and preferences and much depended on the nature of each college's links with employers (Waitt 1980, 402). Much of the provision involved training for craft and technician apprentices.

By the 1950s, the economy started to grow and technical colleges began to expand to meet the needs of local businesses including the provision of apprenticeship training, but it was not until the 1960s that Industrial Training Boards were established with the remit of developing training policies for individual industries and creating syllabuses and qualifications. However, by the 1970s the country's strong industrial base began to decline and with it apprenticeships. Technical colleges began to move away from traditional part-time craft courses into providing more full-time courses including A levels and particularly Business Education Council (BEC) and Technical Education Council (TEC) vocational qualifications. (Doughty, 2015). TEC was founded in 1973 and BEC in 1974. They were responsible for Ordinary National Diplomas and Certificates (OND/Cs) at level 3 and Higher National Certificates and Diplomas (HNC/HNDs) at levels 4 and 5. The diploma qualifications were generally taken on a full-time basis and the certificates on a part-time basis, mainly at technical colleges and the HNC/Ds at polytechnics (Pearson 2023). BEC and TEC offered a wide range of technical and vocational subjects as did a wide range of other examining bodies including the Royal Society of Arts (RSA), City and Guilds of London Institute (CGLI) and The Pitman Institute.

The Emergence of Technical Education

The New Labour Government in 1997 was elected on a landslide which made its pitch to the public as being based on 'education, education, education'. Linked to this was a drive to develop 'skills'. A number of initiatives were put in place to support this—including 'Centres of Vocational Excellence', curriculum reforms, new qualification frameworks and more. The reforms followed the *1997* 'Kennedy Report' and the 'Learning Works: Widening Participation in Further Education', a report by its Widening Participation Committee, chaired by Helena Kennedy QC (Kennedy 1997). This report acknowledged that 'the introduction of competition in publicly funded further education has been accompanied by improved responsiveness to learners' needs', but warned that, as some providers 'competed for those students most likely to succeed', competition also 'inhibited the collaboration needed to widen participation'. Further education needed to 'expand the demand for learning as a whole' (Kennedy 1997, 35). The committee proposed an 'Agenda for Change', which included:

- A government campaign to create a 'learning nation';
- redistribution of public resources 'towards those with less success in earlier learning';
- the establishment of 'a lifetime entitlement to education … which is free for young people and those who are socially and economically deprived';
- the creation of 'a national network of strategic partnerships to identify local need, stimulate demand, respond creatively and promote learning';
- encouraging employers to provide learning centres linked to a 'University for Industry';
- reform of financial support to students to promote equity and 'Welfare to Work through Learning'; and
- the setting of new national learning targets and local targets for participation (Kennedy 1997, 13–14).

This was followed by the 1997 Dearing Review: Higher Education in the learning society which focused on expanding higher education and made way for system of student loans. It recommended 'further reviews of higher education, and stated that the Dearing committee, 'should report every ten years' (Dearing 1997, 356). In 1998, there was a Green Paper 'The Learning Age: a renaissance for a New Britain'. In the forward, Secretary of State David Blunkett wrote that the government's approach to lifelong learning would be based on two key initiatives:

- Individual learning accounts which will enable men and women to take responsibility for their own learning with support from both Government and employers
- The University for Industry which will offer access to a learning network to help people deepen their knowledge, update their skills and gain new ones. (Department for Education and Employment 1998, 8)

A wide range of new bodies were created, including a Training Standards Council, a National Skills Task Force, National Training Organisations, Regional Development Agencies and local learning centres, with greater cooperation between the Further Education Funding Council and local authorities. This led to the 1999 White Paper 'Learning to Succeed'. In it the government set out a vision for the 'Learning Age', in order to build a new culture of learning and aspiration which will 'underpin national competitiveness and personal prosperity, encourage creativity and innovation and help build a more cohesive society'. This particular White Paper states that the government wants 'everyone to benefit from the opportunities that learning brings both in personal growth and the enrichment of communities.... In the digital age, learning must take place 'on-site' in small and medium sized companies as well as in large businesses' (ibid., 13).

While this elevated the role of skills, according to the 2010 skills strategy, England was 'currently weak in the vital intermediate technical skills that are increasingly important as jobs become more highly skilled and technological change accelerates' (Department for Business Innovation and Skills, 2010, 4). The Skills Commission in 2011 found 'the UK had 1,069,000 technicians operating in Science, Technology, Engineering and Maths (STEM), which accounts for 3.7 per cent of the workforce (UK Commission for Employment and Skills 2012). The Dual Mandate consultation paper for Adult Vocational Education (Dept for Business Innovation and Skills 2015) acknowledged that 'higher level vocational training had fallen through the gap between our FE and HE systems... relative to other countries we are behind where we need to be' (DIUS 2008, 7). The paper was commissioned and authored by Vince Cable when he was Secretary of State for Business. He presided over the politically controversial decision to increase HE tuition fees to a maximum of £9000 per year (partly to ensure that following the 2008 financial crash there remained funding for colleges) and in doing so strengthened the market approach to FE and HE which he had extolled through 'Students at the Heart of the System'. The Dual Mandate paper stated (p. 17) that 'we believe it is best to see adult further education as having a dual mandate' and contended that there are two core purposes which adult further education serves:

- 'To provide vocational education for the workplace with a focus on higher level professional and technical skills. Further education colleges were initially developed as civic enterprises by businesses and local authorities, teaching skills that were demanded by employers. This remains the essential core of further education.'
- 'The second part of it is to provide second chances for those who have not succeeded in the school system. The issue here is stark. We estimate that there are 8.1 million adults in England who do not have the numeracy expected of an eleven-year-old child leaving primary school.'

The coherency of the skills landscape may have been curtailed in part by the skills brief shifting between various government departments—business, to education and back again. However, the notion of people gaining higher-level professional and technical education have made their way into the structure of the 2021 Post-16 Education and Skills Bill. While there are qualifications at these levels, the language assumes that the level of study goes beyond the qualification—to demonstrate increasing levels of mastery of a craft. However, the measurement of such education remains embedded in narrow and traditional ways of measuring education. The OECD report links the need for higher-level skills and income and sustainable jobs, and therefore hints at social mobility. If income rates are to increase then people can participate more fully in the society, democracy and contributing to the places they live. While the structure of the learning and skills sector takes on a neoliberal approach in this way the actual funding methodology to achieve this is undergoing social engineering. For example, the Apprenticeship Levy is a part of policy that will bind people to higher and degree apprenticeships through an employer, often with the content being delivered in their local college or university. The Dual Mandate goes on to state:

> Over the course of this Parliament .... we aim to create more opportunities, skilled jobs and to make the UK more competitive so British businesses can thrive and compete with rising economies. The purpose is to shape the system for the long-term giving business the confidence to invest so that as a whole it meets the needs of the modern economy.

It also articulated the identification of five cross-cutting areas of activity which are seen as vital to the overall performance of the economy:

- Sector partnerships: providing support for all sectors to help increase global competitiveness, support innovation and maximise export potential; with tailored strategies for 11 key sectors.
- Technologies: supporting the development and commercialisation of technologies where the UK has the research expertise and business capability to become a world leader.
- Skills: working to deliver the skills that employers need, giving businesses more say over how government funding for skills is spent.
- Access to finance: helping businesses get the finance they need to invest in people and equipment and to grow.
- Procurement: developing UK supply chains and creating a simpler and more transparent public sector procurement system.

Funding Technical Education

While the need for skills development and expansion took hold in political discourse. The funding for it, and therefore a return on investment for the treasury, needed to be considered. The 2019 report into education spending from the Institute for Fiscal Studies (IFS 2019) revealed that further education colleges and sixth forms faced the largest cuts in education in the preceding years. Between 2010–11 and 2018–19, spending per student fell by 12 per cent in real terms in 16–18 colleges and by 23 per cent in school sixth forms. Following on from larger cuts during the 1990s, and lower growth than most other stages of education during the 2000s, further education spending per 16- to 18-year-old is due to be only about 13 per cent greater in 2018–19 than it was about 30 years earlier. Funding is lowest in school sixth forms and sixth-form colleges. In the 2018–19 academic year, the IFS calculated that spending per student was £4800 in sixth-form colleges, £4900 in school sixth forms and £5900 per young person in Further Education colleges. These differences mainly result from a funding system that provides more to vocational and higher-cost courses, as well as to pupils from deprived backgrounds. There are also regional variations which are present because of uplifts given for student postcodes, London weighting and additional support.

The IFS acknowledged that the government has allocated an extra £300 million for further education colleges and sixth forms for 2020–21 (at 2019–20 prices). This led to a real-terms increase in spending per pupil of over 4 per cent in 2020–21, but will still leave spending per student over 7 per cent below its level in 2010–11 in colleges and over 20 per cent below in sixth forms. Indeed, the IFS estimates that fully reversing cuts since 2010–11 would cost a further £1.1 billion over and above existing plans by 2022–23 (£730 million for school sixth forms and £320 million for 16–18 colleges). This increases to about £1.4 billion to ensure that spending on T levels (a new type of qualification being introduced post 16 to create a binary 'academic' and 'technical' qualification divide) is additional to an unchanged level of spending per student. The IFS goes on to argue that total spending on adult education (excluding apprenticeships) has fallen by nearly two-thirds since 2003–04. This combines cuts of 32 per cent up to 2009–10 and 47 per cent between 2009–10 and 2018–19. These cuts have mainly been driven by falls in learner numbers, which have fallen from 4.4 million in 2004–05 to 1.5 million in 2017–18. The IFS notes that the recent Augar review of post-18 education has recommended reversing a range of the cuts to adult education (detailed later in this chapter). The IFS contends that spending on adult education and training is increasingly focused on apprenticeships. Of the £5.3 billion spent on adult education and apprenticeships in 2003–04, about 21 per cent was spent on apprenticeships or work-based learning. By 2018–19, this rose to 54 per cent. Specific spending on apprenticeships rose by 36 per cent in real terms between 2009–10 and 2018–19. Despite this, the government missed its target for 3 million new apprenticeship starts between 2015 and 2020 by a wide margin (Cavaglia et al. 2022). However, an increasing share of individuals on apprenticeships are now on level 3 and above programmes (over 60 per cent to date in 2018–19 as compared with around 40 per cent in 2011–12). These courses generally offer higher economic returns for institutions, employers and individuals. However, since the Higher Education Funding Council for England promoted Higher and

Degree Apprentices in 2016, there have been a number of Higher Education (HE) providers who have also engaged in this work.

The Independent Review of College Financial Oversight by Dame Mary Ney, conducted in October 2019 and published in July 2020, highlighted some systemic issues within the Further Education sector (DfE 2020). She pointed to the current Framework of Funding and Accountability leading to around 10 funding streams for the sector, an inspection regime of around 50 colleges per year (around 80 per cent of which are graded 'good' or better by the Office for Standards in Education (OfSTED)) and highlighted the precarious nature of the sector which leads to frequent cashflow issues. In 2013, the FE Commissioner Office was established to provide independent advice to government ministers on college financial security and between 2015 and 2019 undertook a series of Area Based Reviews for the sector which led to 57 mergers (with 46 of these receiving financial support). Dame Ney concluded that there 'needs to be a shift in emphasis in promoting the strategic role of the sector' (p 9). At the time the report was published, the government committed to launching a Further Education White Paper, and Dame Ney called for a 'National Strategic Vision' for the sector along with 'Local Planning'. This mantra has been taken up by the Commission looking at the Future of Colleges. This has begun by the issuing of a paper on the importance of 'place' and the 'civic role' of Further Education. Given this context, Exley (2020) asks if 'there has ever been a golden age for FE'. He goes on to explore how a range of policy drivers 'from Leitch to Lingfield' have attempted to map one only to fail. As Exley says: 'Colleges may have won their freedom – but the cost has been profound' (p. 1).

This scene setting for FE is important because it establishes that capacity in the sector has clearly been eroded, as has its ability to create a clear sense of its role in improving life chances. Its reductions in income have been at a time when Higher Education activity nationally has increased.

The Role of Employers

As the need for skills education and improvement has taken hold in political discourses, the power around what skills and where has also been an area of debate, discussion and policy making. As is evident through the history of technical education employers have been central to what is taught and why. Indeed, as UKCES (Universities UK 2016) has said:

> Young people should have the opportunity to develop higher-level skills through a system of integrated pathways between the two forms of provision, one that provides a theoretical underpinning to technical knowledge and one that offers the chance for upskilling in line with economic, operational and technological change.

The paper goes on to argue (p. 3) that:

> To target such policy interventions, the sector needs both a clearer and a more granular understanding of the size and content of provision across both further and higher education: what subjects are being offered where, in what modes, and to what extent do these courses dovetail with skills gaps?

It highlights the fact that there is no one definition of a 'graduate job'. Projecting to 2022 (obviously in a time before the 2019–2021 COVID-19 Pandemic), the report deploys four common frameworks for identifying 'graduate' or 'professional' occupations and finds that, under most models, there will be a continued undersupply of graduates, relative to the number of vacancies demanding them. There will also be unmet demand for workers with higher, but not necessarily degree-level, qualifications, such as BTECs or HNDs. It also points to clear evidence of occupational upskilling, wherein occupational functions, and their attendant skill requirements, rise over time, suggesting that the UKCES analysis may actually underestimate future demand for graduate talent.

It says that:

> Graduates in every discipline pick up a diverse array of 'employability skills', including problem-solving, critical analysis and entrepreneurial skills. However, the provision of these is not, as far as we can tell, evenly spread. For example, creative arts students report that through the course of their university education they developed high-level entrepreneurial skills but not necessarily logical thinking, and maths students report that they developed high-level problem-solving skills but not necessarily presentational skills. The provision of these skills is important, not only to graduates, but to the wider economy. Our analysis may actually underestimate future demand for graduate talent.

The report goes on to make six recommendations which focus on a skills mismatch. However, the main focus is on institutions and employers jointly developing a 'skills translation' exercise to help all parties understand how and where these 'soft skill' principles can be practically developed and applied.

With all of the focus on 'skills', especially 'higher level skills', it may be prudent to try to establish a definition of them. Green and Henseke (2014) define them as: 'knowledge and attitudes as well as technical capabilities'. Of course, assuring 'higher-level' skills is a somewhat trickier enterprise than defining them. We might think of higher-level skills in terms of:

- Qualification level (e.g. Bachelor's Degree, Foundation Degree, GCSE)
- Specific core and employability skills (e.g. literacy, numeracy, problem-solving and leadership)
- Domain or subject-based knowledge

There is little that can be found in the way of policy work that accurately describes how and if these definitions work across Further and Higher Education, yet they formed the basis of government spending in the sectors for teaching activity at the pre-2021 Comprehensive Spending Review (CSR). Moreover, there has been little done by policy makers to try to make sense of the impact of these approaches and what else could be done in future policy activity. For example, following the 2015 General Election, the Conservatives gained an overall majority. This delayed the implementation of the Dual Mandate which had been fronted by the Liberal Democrats. It meant that there was further faith placed in the market to ensure that Higher Education was regulated. The increased marketization led to new policies.

The Apprenticeship Market

Some of the most significant crossovers for collaboration and competition in recent times have emerged in the shape of higher and degree apprenticeships. In 2011, Lord Sainsbury undertook a review into apprenticeship provision and noted that apprenticeship provision was of poor quality (Sainsbury 2016). At the same time, a number of academics, for example Fuller and Unwin (2011), Minton and Lowe (2019) and Crawford-Lee and Morwood (2019) were focusing on comparisons between apprenticeship and skills provision in the UK compared to the rest of the world and were noting a decline in the prestige of apprenticeship provision in

England. However, they noted that it did exist in countries such as Germany where productivity post pandemic was much higher than that of the UK. Sainsbury, through his Gatsby Foundation, put forward a series of reforms to apprenticeships, including the establishment of an Institute for apprenticeships and the changing of the qualification structure so that employees were involved in creating apprenticeship standards focused on the required knowledge skills and behaviours of employers.

Furthermore, Sainsbury argued that apprenticeship assessment was not robust and so sought to establish 'Endpoint Assessment' (that drew the various elements of the apprenticeship provision together). International comparators were also considered. For example, in Germany an 'advanced apprenticeship' was articulated at level six or the equivalent of an honours degree. In England, an 'advanced apprenticeship' was equivalent to a level 3 qualification. This gave rise to the notion that apprenticeship provision should exist at higher levels and Sainsbury advocated reforms around higher and degree apprenticeships with a degree apprenticeship actually having a degree (be that an honours degree or a master's degree) contained within it. At the time of this work, the OfS did not exist, and so, its forerunner, the Higher Education Funding Council for England (HEFCE) was able to establish a 'catalyst competition' to enable providers to build capacity and processes around higher and degree apprenticeships. This meant that the colleges had to collaborate with a university, or colleges and universities had to go into direct competition for the same apprentices. Again, little research has been undertaken as to how Further and Higher Education could work together to respond to the challenges and opportunities policy is attempting to resolve.

Indeed, in its 2020 report, 'Apprenticeships and Social Mobility Fulfilling Potential', the Social Mobility Commission states that apprenticeships "can support employability and enable individuals to gain skills in a non-academic context" (Social Mobility Commission 2020). While this description is not entirely wrong, it is inaccurate in several keyways and does not entirely reflect the Government's definition and policy objectives. Apprenticeship is not focused on just supporting employability; its focus is on ensuring that each individual who completes the programme is occupationally competent. Also, apprenticeship is not about enabling individuals to "gain skills in a non-academic context". The University Vocational Awards Council (UVAC) states:

> We reject a blinkered approach that focuses on social mobility measures in isolation. Instead, we argue that recommendations for supporting social mobility and levelling up must be determined in the context of the role and purpose of Apprenticeship in increasing productivity, supporting diversity in the workforce and enabling our public sector employers to recruit and train the nurses, police constables and social workers they need. The role of Apprenticeship in the Levelling Up Agenda should also be supported nationally and developed locally in this context. (Anderson and Crawford Lee 2020)

The role of apprenticeships in social mobility has also not been easy or necessarily considered by policy makers. More recently it has started to be considered to be a front for widening participation, but it is missed from the OfS Regulatory Framework which has Access and Participation as the first element of Registration as a HE provider. Indeed, early Individualised Learner Record (ILR) and HE Statistics Agency (HESA) data demonstrated that enrolments onto Higher and Degree Apprenticeships (HADAs) from the early days were undertaken by those who came from more affluent backgrounds, were not on free school meals and not from POLAR Quintiles or Index of Multiple Deprivation (IMD) quintiles 1 and 2 (Cavaglia et al. 2022). UVAC's reference to the 'levelling-up agenda' had been a focus of the Borris Johnson government following a landslide election win in December 2019. Following the 2016 BREXIT referendum Johnson had promised the, often Labour voting Northern City Regions, that policies would be put in place to improve their social mobility. From an education perspective, some of the seeds of access to education and a reduction in a market approach had been sown some time previously.

The Augar Review—2019

It was often discussed in newspapers and on podcasts (such as *Rock and Roll Politics*) that during the General Election of 2017 an increased number of young people engaged in politics for the first time. This led to a debate about the increases in tuition fees for higher education study. The outcome of the election was a hung parliament. However, it meant that the government in charge had to do something to be seen as proactive in relation to looking at the landscape of post-18 learning. Consequently, in February 2018, the Prime Minister, Teresa May, announced a wide-ranging Review of Post-18 Education and Funding led by Philip Augar (Augar 2019). The Review was partly in response to increased debate around the cost and value of higher education following a period of reform which saw tuition fees rise to £9250 per year, maintenance grants abolished, and typical student debt rise to £47,000 for a three-year degree. During the period of higher education reforms, the further education sector had also experienced financial difficulties as the IFS data above shows. The Review therefore aimed to create a joined-up post-18 education that would work for students and taxpayers. The Review panel was made up of five experts from the HE and FE sector: Dr Philip Augar, Professor Ivor Crewe, Jacqueline de Rojas CBE, Professor Edward Peck, Beverley Robinson OBE and Professor the Baroness Alison Wolf.

The Review report was published on 30 May 2019 and it gave a detailed analysis of the post-18 education sector and the funding issues faced by stakeholders. The report contained 53 recommendations on the future structure of the sector and funding proposals. The headline recommendations are:

- The reduction of higher education tuition fees to £7500 per year
- Government to replace lost fee income by increasing teaching grant
- Extending the student loan repayment period from 30 years to 40 years
- Reducing the interest charged on student loans while students are studying
- Capping the overall amount of repayments on student loans to 1.2 times their loan
- Reducing the income threshold for student loan repayments from £25,000 to £23,000
- Reintroducing maintenance grants of £3000 for disadvantaged students
- Introducing maintenance support for level 4 and 5 qualifications
- A first free full level 2 and 3 qualification for all learners

The report acknowledges that post-18 education in England is a "story of both care and neglect" and it proposes that the HE sector should absorb a further freeze on per student resources to help fund investment in other parts of the post-18 education system. This would put state subsidy back into the FE and HE education systems and thus limit market forces. The proposals would cost an additional £0.3–0.6 billion in annual ongoing annual costs plus a one-off £1.0 billion on capital for further education colleges. The costs arise from extending entitlement to maintenance and tuition support to level 4 and 5 qualifications and extending level 2 and 3 entitlements. The changes to student finance and funding are expected to reduce costs when taken on their own. They shift the balance of taxpayer funding from loan write-offs to more direct funding for teaching and maintenance. In response to the report, Debbie McVitty, editor of the HE sector body 'WONKHE' said:

> Having been given permission in the terms of reference to address the whole post-compulsory education system the review is quietly radical in its inclusion of both higher and further education in its definition of the mission of the post-compulsory education system to undertake research, scholarship and innovation, promote the ability of citizens to reach their full potential, contribute to civic wellbeing and meet the economic and skills needs of the country. (McVitty 2019, 1)

The Johnson government that inherited the report after the fall of the May government, had not published a response by May 2021. However, there was a series of White Papers and Bills which seemed to slowly and incrementally introduce Augar and build on the Dual Mandate. The direction of travel was set by the administration considering whether or not to reduce funding for arts-based subjects and promoting more students into STEM (Science, Technology, Engineering and Maths) subjects. However, as Rich noted on the WOKHE Podcast on May 29th 2021, all those that want to enter STEM subjects probably can, and so the question might be how employers could have an enhanced role to contribute to the system. It does appear that the role of employers will make a greater contribution to the future shape of FE.

Skills White Paper and Subsequent Skills Act

In March 2021, the government introduced a 'Skills and Technical Education' White Paper which was followed by a Bill announced in the Queen's Speech of May 2021 (DfE 2021). This became an Act in 2022, (DfE 2022). It intended to herald strategic, systemic change and work across government for the whole skills sector. However, it majored on FE. The aim was to legitimise the work of the FE and skills sector by legislating to create 'local skills plans' where employers dictate local skills needs. There is, however, little consideration of how this maps into the work of devolved authorities or of practices that currently take place or other means for employers to contribute to the skills system. Additionally, the Act includes a Lifelong Learning Loan Function, which focuses on FE funding, not on HE funding, and lacks clarity as to how the two elements connect. Within the Act Institutes of Technology are deliberately established as HE and FE regional collaborations along with a higher-level skills sector and an increased regulatory role in this for Institute for Apprenticeships and Technical Education. There is increased scope for the Secretary of State to use powers to the regulate provision, allowing for a more planned and less demand driven approach to the skills market, in addition to the FE Commissioners intervention regime, Additionally, the Act states that by 2030, the DfE will have aligned the 'vast majority' of post-16 technical education to employer standards (which are predominantly contributed to by larger employers from larger conurbations); The Act establishes the Higher Level Skills and Productivity Board to advise the Secretary of State on this work from the perspective of economists. However, for the first time since the Dual Mandate, Higher Level Technical and Professional education becomes part of mainstream political thinking.

Higher Level Technical Education

In his report to the Gatsby Foundation in 2019, Simon Field articulated how 'England is distinctive in its lack of mid-level education and training' (Field 2018, 5). Data drawn from 2014/2015 demonstrates that around 190,000 individuals were enrolled on level 4 or 5 qualifications—which represents about 4 per cent of higher education provision (in Scotland for example it was 18 per cent). In New Zealand the number of graduates at this level in the same year was nearly 750,000, and in Germany it was nearly 300,000. Field contended that this lack of 'middle level' education was a weakness. He argues that the jobs market needs more than people who have a three-year bachelor's degree and asserts 'the one clear certainty about the future is that fast-changing labour market requirements and short targeted technical programmes are vital elements in the response' (p. 5). In 1945, the main Higher Level Technical Qualifications were Higher National Certificates (HNCs) and Higher National Diplomas (HNDs) which had been formed in the 1920s through a collaboration between the Board of Education and Professional Bodies.

Following the 1945 Percy Report, three types of college emerged—regional / local, evening and Colleges of Advanced Technology (CATs) (Percy 1945). These institutes delivered a 'Dip. Tech' qualification that was the equivalent to a bachelor's degree and involved a 'sandwich' industry placement element. This approach to work with industry has continued into the traditional 'sandwich year' that is known in HE today. The 1963 Robbins proposal was a 'milestone' (Robbins 1963, 6). It proposed a big expansion in HE, backed by extensive public funding. Growth was focused on universities and colleges that would later become known as 'polytechnics'. Consequently, there was great emphasis placed on the three-year bachelor's degree programmes as opposed to level 4/5 technical education. Crosland's original 1965 proposal for polytechnics gave them a higher-level technical vision and mission (Crosland, 1965). By the 1989/1990 academic year there were 280,000 students in English Polytechnics, and, as detailed, by 1992 Polytechnics became Universities. Alongside this there were changes to qualifications. The Technical Education Council (TEC) and the Business Education Council (BEC) were established in 1973 to create a more systematic framework for technician education, and they merged in 1983. They focused on qualifications such as HNCs and HNDs. In 1997, the Dearing Report, while also proposing tuition fees for 'traditional' HE, proposed that the next expansion in Higher Education should be HNCs and HNDs. This led to the launch of Foundation Degrees and Foundation Degree Forward (an agency to support the development of Foundation Degrees) in the Year 2000. Indeed, the significant growth of Foundation Degrees was validated or franchised by Higher Education providers and delivered by Further Education Colleges. This then led to a sequence of other initiators which emerged from the Post-16 Skills Plan in 2016 which was put in place after the Sainsbury Review (detailed above). From here there has been the emergence of T Levels (which are legislated for in the Skills Act).

What did surprise government, however, was the reduction in part-time HE as a result of the Brown Review into HE funding. Part-time provision halved in size between 2008/2009 and 2015/2016. In 2017, Nick Hillman, Director of HEPI, wrote:

> The collapse in part-time study is arguably the single biggest problem facing higher education at the moment. There are other challenges too, such as the future of the research environment, how to assess the quality of teaching and dealing with the effects of marketisation. But it is the fall in part-time

> learning that is probably the biggest black spot. The fall in part-time student numbers is clearly partly – possibly mainly – associated with the changes to student finance, hence the title of this collection. Part-time numbers have fallen more in England than other parts of the UK with lower (or no) fees, but it is not the sole cause. The decline began before the £9000 fees were introduced. Any solution is likely to rest upon innovative delivery methods and other ways of improving access as much as relying on tweaks to the entitlement for financial support. If we succeed in reversing the decline in part-time study, the benefits to employers in terms of improved productivity and to the economy in terms of faster growth will be substantial. But the benefits to individuals and their families will be even more transformative. Lifelong learning is a concept that no one opposes, but it does not happen on its own and it needs to be supported. (p. 2)

At frequent intervals the government has expressed concern about the need for better technical education. In 1961, a White Paper 'Better Opportunities in Technical Education' proposing a series of recommendations which included the introduction of new courses for technicians to be called T level courses (Evans 2022b). These concerns were raised again with the Review of Vocational Education—The Wolf Report (Wolf 2011), which suggested that vocational education was no longer effective and made a number of recommendation including that educational institutions would be able to offer any vocational qualification they wished, that students who are under 19 should continue to study to ensure that they have GCSE A-C in English and Maths, that students should be able to move sideways or even down in order to change subject or sector and that the apprenticeship system should be reviewed (Wolf 2011, 14–15). This was followed by the Sainsbury Report (Sainsbury 2016), which reiterated that there was a shortage of young people trained as technicians and that the existing system is over-complex and does not meet the needs of employers. The report suggested that the size of the post-secondary technical education sector in England is extremely small by international standards. This adversely affects the UKs productivity, where at that time we lagged behind competitors like Germany and France by as much as 36 percentage points (p. 8).

T Levels

History

In Chap. 1, the history of technical education in England was discussed. The origins can be traced back to the mediaeval guilds from about 1250 where part of the role of the master was to educate and train the apprentice in the craft (Evans 2022). In each century, the need for technical education has continued, including in 1571 when a scheme was established in Norwich to teach techniques in the manufacture of clothing, in 1663 when the Royal Society was established to improve natural knowledge and in 1760 the Mechanics Institutes movement began (Evans 2022a). However, it was only with the Technical Instruction Act 1889 that the government realised that to compete with other European countries state involvement was needed (Green 1995). Under this Act, local committees were formed to create institutions that could provide technical education. The Education Act 1902 created Local Education Authorities (LEAs) who administered schooling including technical education, but the law was permissive and thus led to a wide variation in provision. The Education Act 1944 created a formal further education sector was created by the Education Act 1944 which tasked LEAs with providing adequate full and part-time provision for post school-age students and a shift began from evening study into full-time study and the creation of more technical colleges. However, the provision differed markedly depending upon each LEA's priorities and preferences and much depended on the nature of each college's links with employers (Waitt 1980, 402). Much of the provision involved training for craft and technician apprentices.

After the Second World War and by the 1950s, the economy started to grow and technical colleges began to expand to meet the needs of local businesses including the provision of apprenticeship training but it was not until the 1960s that Industrial Training Boards were established with the remit of developing training policies for individual industries and creating syllabuses and qualifications. However, by the 1970s the country's strong industrial base began to decline and with it apprenticeships. Technical colleges began to move away from traditional part-time craft courses into providing more full-time courses including A levels and particularly Business Education Council (BEC) and Technical Education Council (TEC) vocational qualifications. (Doughty 2015). TEC was founded in 1973 and BEC in 1974. They were responsible for Ordinary Diplomas and Certificates (OND/Cs) at level 3 and Higher National Certificates and Diplomas (HNC/HNDs) at level 4 and 5. The diploma qualifications were generally taken on a full-time basis and the certificates on a part-time basis, mainly at technical colleges and the HNC/Ds at polytechnics. (Pearson 2023). BEC and TEC offered a wide range of technical and vocational subjects as did a wide range of other examining bodies including the Royal Society of Arts (RSA), City and Guilds of London Institute (CGLI) and The Pitman Institute.

At frequent intervals, the government has expressed concern about the need for better technical education. In 1961, a White Paper 'Better Opportunities in Technical Education' proposing a series of recommendations which included the introduction of new courses for technicians to be called T courses (Evans 2022b). These concerns were raised again with the Review of Vocational Education—The Wolf Report (Wolf 2011), which suggested that vocational education was no longer effective and made a number of recommendation including that educational institutions would be able to offer any vocational qualification they wished, that students who are under 19 should continue to study to ensure that they have GCSE A-C in English and Maths, that students should be able to move sideways or even down in order to change subject or sector and that the apprenticeship system should be reviewed (Wolf 2011, 14–15). This was followed by the Sainsbury Report (Sainsbury 2016), which reiterated that there was a shortage of young people trained as technicians and that the existing system is over-complex and does not meet the needs of employers. The report suggested that the size of the post-secondary technical education sector in England is extremely small by international standards. This adversely affects the UK's productivity, where at that time the country lagged behind competitors like Germany and France by as much as 36 percentage points (p8). The Sainsbury Report suggested a number of recommendations.

Introduction of T levels

The government accepted the Sainsbury recommendations (DFE 2016). In this skills plan the government suggested that in future there would be two routes of education post 16, the academic A level route and a technical route including apprenticeships which would have the same standing as A levels. The qualifications in the technical route would fall into 15 occupational routes, be employer led and be overseen by an Institute for

Apprenticeships. The next stage was to introduce the new technical qualifications to be called T levels (DfE 2017). T levels would initially be introduced at level 3 and would replace existing technical level provision. They would be designed primarily to support progression to skilled employment although they would also be a route to higher qualifications. Both T levels and apprenticeships would be based on the same standards, but the input would change depending whether they were T levels being delivered on a full-time basis at a college or school or apprenticeships which would be delivered mainly in the workplace. T levels would be equivalent to A levels and have more teaching time than the current vocational qualifications. All T levels programmes would have a substantial period of work placement. A limited cohort of three of the routes would be piloted in 2020 and all routes would be ready to be delivered by 2022. The standards and assessment plans would be produced by the Institute for Apprenticeships and panels of employers will be responsible for producing the outline content of the T levels. Training in how to deliver the T levels would be provided for teachers.

There then came a series of action plans in 2018, 2019, 2020 and 2022/23 as T levels were gradually introduced although they were not introduced as quickly as the government first announced. However, by 2021 students had enrolled on courses for the first five T levels in the Digital, Construction and the Built Environment, Health and Science, and Education and Early Years routes (DfE 2023). A further three routes in Business and Administration, Legal, Finance and Accounting and Engineering and Manufacturing had begun in September 2022. At that point in March 2023, four more routes were planned for September 2023, six for 2024, two for 2025 and two more as some point after 2024. Thus the 15 routes originally planned had now expanded to 24 and many routes are broken down further. For instance, the digital route had now expanded to Digital Support Services and Digital Business Services.

The first three T levels were delivered by 43 providers and in Summer 2022, 1029 students gained the qualification (DfE 2023, 11). In 2021, 10 T Levels were available, and 102 providers were delivering, with 5210 new students starting the programme. By 2022, there were 16 T Levels with 164 providers who had recruited around 10,200 students. In 2022, the most popular route was Education and Early Years with a cumulative total of 2150 students followed by 1800 studying Health and Science and Digital with 1750 students. Almost all of these students had GCSE grade 4 or higher in both English and Maths.

What Is a T Level?

As discussed above, T levels are a vocational alternative to A levels which are designed to have the same standing as A levels. Each T level is made up of a number of components. The Core Component is made up of two parts, a core examination which contributes between 60 and 75 per cent of the weight towards the core and an employer set project which makes up the remainder. The other part is the occupational specialism assessment which is an externally set assignment. If a student does not have level 4 GCSE in English and maths they must continue to work towards a level 2 qualification in those subjects (City and Guilds 2023). Technically, applicants for T levels do not need level 4 GCSE's in English and maths to take a T level but most colleges and schools require them. In addition, students must undertake a period of work placement which should be relevant to the occupational specialism. Initially, when T levels were introduced, the placement had to be with one employer although this has now been changed to two employers if necessary and the number of hours is a minimum of 315 although now up to 35 hours can be work taster activities. If a student does not meet these requirements, they cannot receive a T level certificate (Department for Education 2023a). The final grade that a student receives will be a pass, merit, distinction or distinction*. The grade is made up of two components, a grade A*-E for the core component and a grade for the occupational specialism of pass, merit or distinction. They are combined to create the overall grade (Department for Education 2023b).

Students who wish to take a T level but are not judged by providers to be ready can take a T level Transition Programme. These programmes are designed to equip a learner with the necessary knowledge and skills to enable them to progress on to T level. They are made up of the following components:

- Industry relevant technical knowledge and skills
- Skills for successful study
- English, maths and digital skills
- Knowledge and skills for the workplace
- Positive attitudes and behaviour

How the students will be assessed is decided by the provider but in line with T level assessment methods and a final certificate will not be awarded although providers can issue their own (Department for Education 2021).

Controversy

As mentioned above, the Post-16 Skills Plan of 2016 (Department for Education 2016) accepted the Sainsbury recommendations and indicated that a thorough review of post-16 technical education needed to take place because the current system was flawed. The existing technical qualifications had not always been produced in conjunction with employers, there were too many overlapping and low value qualifications, the system was complex and difficult to navigate, there are not enough apprenticeship opportunities, and programmes are not necessarily designed to deliver what is needed to move to skilled employment. Consequently, the plan was to move to a much simpler system where young people at the age of 16 could choose to take A levels, or a T level or an apprenticeship. The T levels and apprenticeship content would be decided in conjunction with employers and the Institute for Apprenticeships would ensure high-quality provision. Most other level 3 qualifications would have funding withdrawn.

This caused concern in many quarters including the House of Lords. The Protect Student Choice organisation was established to object to the defunding of applied general qualifications such as BTECs. The organisation has 29 members which include The University Alliance, Million Plus (which represents modern universities, Universities UK, The Independent Schools Council, UNISON and the NUS (Protect Student Choice 2023). The Protect Student Choice organisation argue that:

- The current three route model with A levels, technical qualifications that lead to a specific occupation and applied general qualifications such as BTECS works well and should be retained.
- T levels and a different type of qualification which should sit alongside the existing qualifications.
- Removing funding from some BTECs will leave some students from Asian and disadvantaged backgrounds and those with special educational needs with no progression pathway at the age of 16.
- Defunding BTECs will reduce participation in higher education with disadvantaged groups.
- The government has used misrepresented date to predict the uptake of T levels (Lewis and Wilson 2023, 6).

Moreover, in April 2023, The House of Commons Education Committee called for a moratorium on the defunding of applied general qualifications until there is a robust evidence base proving that T Levels are demonstrably more effective in preparing students for progression, meeting industry needs and promoting social mobility (Lewis and Wilson 2023, 6). It was reported in January 2022 that a number of universities refused to accept T levels including ten Russell Group universities as they were unsure whether T levels would meet their criteria for entry. However, the number is gradually increasing (Camden 2022).

The government indicated that funding would cease for 134 qualifications from 2024 which overlap with the T levels in Digital, Construction, Education and Healthcare. They also announced in May 2023 that funding will cease for qualifications which overlap with T levels in business and administration, engineering and manufacturing, and finance and accounting. One of the courses that will be affected included the BTEC National Foundation Diploma in Engineering and an equalities impact assessment was produced that removing qualifications such as this could have an adverse impact on male students, black students, students with special educational needs, students with Education, Health and Care plans and students from disadvantaged backgrounds (Lewis and Wilson, 2023, 22). The Social Market Foundation found that students accepted to university from working-class backgrounds and/or minority ethnic backgrounds are more likely to hold a BTEC qualification than their peers (Gicheva and Petrie 2018).

As seen above, one of the major concerns about T levels is that they are more academic in nature than the vocational programmes they are replacing in the content, the way they are taught and the way they are assessed. Thus, a major concern is that a whole socio-economic tranche of young people will be disenfranchised. Navelles (2022) argues that this will particularly impact on young people in rural areas. She suggests that because T levels are likely to be less attractive than the courses, they replace fewer schools and colleges are likely to offer a full range of T levels and students from poorer backgrounds will be unable to pay the fares to travel to distant colleges or schools. She understands that the construction route is only going to be offered by one college in Somerset and two in Gloucestershire.

Remedies

To enable young people to access T levels who may not be fully prepared or be able, they can take the T level transition programme which is a one-year programme which prepares students for taking a T level (Department for Education 2023b). This programme involves studying technical content aligned to T Levels while developing English, maths and digital skills, with some work experience and personal development. Although technically entry to T level does not require any level of GCSE, in reality most colleges will expect at least 4 GCSEs including English and Maths at grade 4. Thus, the transition programme also gives students the opportunity to improve their English and maths grades. This programme does not lead to a qualification although the Level 2 BTEC programmes it notionally replaces do.

Currently, many schools and some colleges combine vocational qualifications such as BTEC Nationals with A levels because Nationals can be taken at three levels which equate to one or two A levels. Consequently from 2025, small alternative academic qualifications will be available that are up to the size of an A Level and will be taken in conjunction with A Levels (Lewis and Wilson 2023). They will 'complement and enhance the A level offer', and be available in 'strategically important' subjects, such as science, technology, engineering, maths, and those supporting the NHS, as well as subjects less well-served by A Levels. In addition, large alternative academic qualifications will be introduced which will be at least the size of two A Levels and generally be taken on their own. They will be available in subjects where there are no T Levels, there is a need for a large qualification enabling entry to more specialist areas of higher education, such as performing arts, and there is a clear and direct progression link into higher education.

Case Study

The writer has met with tutors, heads of departments and senior managers from five further education colleges to ascertain their views on the T levels they are delivering or are about to deliver. Three of the colleges has been delivering T levels for a while, while the other two were just starting. They all indicated that they had done a great deal of planning and development and had been frustrated by the frequent changes made by the D of E. The consensus was that T levels can be very good for the brightest students who would otherwise have taken A levels because the style of teaching needed for T levels and the assessment regime is more similar to A levels than the vocational courses which T levels are replacing. The previous vocational programmes such as BTEC Extended Diplomas had some external assessment but most of the programmes are assessed by assignments which are marked by tutors. The assignments are usually vocationally based case studies and used as a method of learning. Although the tutor may deliver some of the knowledge that is required to complete an assignment the onus is upon the student to research and gather information themselves. Some assignments may be produced by a team of students and others require student to give presentation of their findings. Some young people find they can perform better and are more comfortable with this assignment-based approach rather than the didactic, mainly exam-based approach of A levels. The staff I met at the college felt that the assignment-based approach was often more attractive to students from lower socio-economic backgrounds, and this is supported by research (Dilnot et al. 2022).

Consequently, in all the colleges they initially ran all the T level programmes alongside the existing BTEC Diplomas as they were concerned that many of the students who can successfully complete a BTEC programme will not be able to successfully complete a T level. Consequently, the numbers on T level programmes were generally very low with an average of six or seven on most programmes compared with the BTEC programmes with an average of 20. Managers were concerned that as the BTEC programmes are defunded the students who will then have to take the T level will perform poorly.

Managers were also concerned about the difficulty in obtaining the extensive work experience required for T levels, particularly in the rural areas surrounding the colleges and although the DfE had gradually reduced the requirements for work placements, they believed many students would not be able to find appropriate placements. This applied particularly to students from lower income families in rural areas who would not be able to travel to the cities or large towns where placement were more likely to be available. One group of colleges who had previous offered the BTEC Health Diploma which enabled students to cover all aspects of nursing, midwifery and social work now found that they could only offer the Adult Nursing T level stream because hospitals would not accept students for child nursing or midwifery because they were under 18 years of age and also there is not a T level stream for those interested in social work as there was previously. However, it was pointed out that those very bright students who took the adult nursing T level got a much better work experience than they would have had on the BTEC programme. Similarly, with the small numbers of brighter students in groups taking the

Digital and Business and Management T levels the students were able to undertake very good work experiences.

A manager responsible for the construction dept in one of the colleges pointed out that students who take the construction T level who are interested in bricklaying take eight areas in the occupational part of the course but only one of those actually involves laying bricks. Thus, he could not see how any of the students who we currently taking the BTEC diploma would want to take the T level.

T Level Conclusion

Ofsted, the inspecting body for schools and FE colleges, have been carrying out thematic reviews of T levels and in their 2023 review they came to a number of conclusions (Ofsted 2023). They found that the experience of students on T levels differs markedly based on the college they attend. They also found that the awarding bodies have been slow in producing teaching materials and many tutors have difficulty in adapting to the academic nature and depth of T levels. Teachers expressed concern about the high volume of content on T levels and the time they had to deliver it which is eroded by lengthy assessment and work placements. They found that many students do not progress on to the second year of the course. Inspectors found that the quality of work placements differed widely and that colleges had difficulty in finding appropriate placements, particularly in rural areas.

At the time of writing, it is expected that T levels will continue to be rolled out and other vocational qualifications will be defunded although the Labour shadow government have indicated that if they come into power they will pause and review the removal of funding (Lewis and Wilson 2023, 7). The government have announced the introduction of an Advanced British Standard which will replace both A levels and T levels with a broader curriculum but will not come into effect until about 2033.

Summary

This chapter has examined how government policy and a series of reports have influenced the role that technical education plays, the role of the education bodies who deliver it and the impact that policy has had on learners.

References

Anderson, A., and M. Crawford Lee. 2020. *Realising the Potential of Higher and Degree Apprenticeship in Supporting Social Mobility and the Levelling Up Agenda*. UVAC.

Augar, P. 2019. Review of Post 18 Education and Funding, DofE. Available in https://www.gov.uk/government/publications/post-18-review-of-education-and-funding-independent-panel-report (Accessed 27 May 2024).

Camden, B. 2022. *Confused and Frustrated, Most Universities Reject First Cohort of T Level Students*. London: FE Week.

Cavaglia, C., S. McNally, and G. Ventura. 2022. *Evolution of Apprenticeships—Apprenticeship Pathways and Participation Since 2015*. London: Sutton Trust.

City and Guilds. 2023. *T Level Qualifications, Teaching, Learning and Assessment Guide*. London: City and Guilds.

Crawford-Lee, M., and S. Morwood. 2019. Degree Apprenticeships: Delivering Quality and Social Mobility? *Higher Education, Skills and Work-Based Learning* 9 (2): 134–140.

Crosland, A. 1965. *Woolwich Polytechnic Speech*. DfES.

Dearing Report. 1997. *Higher Education in the Learning Society Main Report*. London: Her Majesty's Stationery Office.

Department for Business Information and Skills. 2010. *Skills for Sustainable Growth*. DfBIS. Available in https://www.gov.uk/government/publications/skills-for-sustainable-growth-strategy-document (Accessed 27 May 2024).

Department for Business Innovation and Skills. 2015. *A Dual Mandate for Adult Vocational Education*. DfBIS. Available in https://assets.publishing.service.gov.uk/media/5a7493dc40f0b61938c7e945/bis-15-145-A-dual-mandate-for-adult-vocational-education.pdf (Accessed 27 May 2024).

Department for Education. 2016. *Post 16 Skills Plan*. DfE. Available in https://assets.publishing.service.gov.uk/media/5a80d94ded915d74e6230cbe/Post-16_Skills_Plan.pdf (Accessed 27 May 2024).

———. 2017. *Post-16 Technical Education Reforms T Level Action Plan*. DfE. Available in https://www.gov.uk/government/publications/t-level-action-plan (Accessed 27 May 2024).

———. 2020. *Report of the Independent Review of College Financial Oversight*. DfE. Available in https://www.gov.uk/government/publications/report-of-the-independent-review-of-college-financial-oversight (Accessed 27 May 2024).

———. 2021. *Skills for Jobs: Lifelong Learning for Opportunity and Growth*. DfE. Available in https://www.gov.uk/government/publications/skills-for-jobs-lifelong-learning-for-opportunity-and-growth (Accessed 27 May 2024).

———. 2022. *Skills and Post 16 Education Act*. DfE.

———. 2023. *T Level Action Plan 2022 to 2023*. DfE. Available in https://assets.publishing.service.gov.uk/media/6627c087d29479e036a7e68e/T_Level_Action_Plan_2023_to_2024.pdf (Accessed 28 May 2024).

———. 2023a. *T Level Industry Placements: Delivery Guidance*. DfE. Available in https://www.gov.uk/government/publications/t-level-industry-placements-delivery-guidance (Accessed 28 May 2024).

———. 2023b. *Introduction of T Levels*. DfE. Available in https://www.gov.uk/government/publications/introduction-of-t-levels/introduction-of-t-levels (Accessed 28 September 2024).

Department for Education and Employment. 1998. *The Learning Age a Renaissance for a New Britain*. DfEE.

Dilnot, C., L. Macmillan, and G. Wyness. 2022. *Educational Choices at 16–19 and University Outcomes*. Oxford: Nuffield Foundation.

DIUS. 2008. *Departmental Report*. Department for Innovation, Universities and Skills.

Doughty, R. 2015. *70 Years of Change and Challenge in FE colleges—1945–1992*. AOC Jobs. 11 September 2015.

Evans, R. 2022. *A Short History of Technical Education—Chapter 3, The Guilds and Apprenticeships*. https://technicaleducationmatters.org/2009/05/13/chapter-3-the-guilds-and-apprenticeships (Accessed 28 March 2023).

———. 2022a. *A Short History of Technical Education—A Chronology, Technical Education Matters*. https://technicaleducationmatters.org/category/publications/publications-ashte (Accessed 28 March 2023).

———. 2022b. *Technical Education Matters, Chapter 5—Developments in the 20th Century*. https://technicaleducationmatters.org/2009/08/22/chapter-5-developments-in-the-20th-century-continued (Accessed 29 March 2023).

Exley, S. 2020. "Set It Down with Gold on Lasting Pillars": In Search of FE's Golden Age. *FE News*. https://www.fenews.co.uk/fe-voices/set-it-down-with-gold-on-lasting-pillars-in-search-of-fe-s-golden-age/ (Accessed 14 April 2024).

Field, S. 2018. *The Missing Middle: Higher Technical Education in England*. Gatsby Foundation.

Fuller, A., and L. Unwin. 2011. Apprenticeship as an Evolving Model of Learning. *Journal of Vocational Education and Training* 63: 261–266.

Gicheva, N., and K. Petrie. 2018. *Vocation, Vocation, Vocation: The Role of Vocational Routes into Higher Education*. London: Social Market Foundation.

Green, A. 1995. Technical Education and State Formation in Nineteenth–Century England and France. *History of Education* 24 (2): 123–139.

Green, F., and G. Henseke. 2014. *Graduate Jobs in OECD Countries: Development and Analysis of a Modern Skills-Based Indicator*. Institute of Education.

Hillman, N. 2017. *It's the Finance, Stupid! The Decline of Part-time Higher Education and What To Do About It*. HEPI.

Institute for Fiscal Studies. 2019. *2019 Annual Report on Education Spending in England*. IFS. Available in https://ifs.org.uk/publications/2019-annual-report-education-spending-england#:~:text=Education%20spending%20is%20the%20second,about%204.2%25%20of%20national%20income (Accessed 28 May 2024).

Kennedy, H. 1997. *Learning Works Widening Participation in Further Education*. Further Education Funding Council: Coventry.

Lewis, J., and S. Wilson. 2023. *The Reform of Level 3 Qualifications in England*. London: The House of Commons Library.

McVitty, D. 2019. *The Augar Review: The Essential Overview for HE*. WONKHE.

Minton, A., and J. Lowe. 2019. How Are Universities Supporting Employers to Facilitate Effective 'On the Job' Learning for Apprentices? *Higher Education, Skills and Work-Based Learning* 9 (2): 200–210.

Navelles, G. 2022. *Levelling Up Skills: Are T Levels Good for Social Mobility?* Business West. https://www.businesswest.co.uk/blog/levelling-up-skills-are-t-levels-good-social-mobility (Accessed 8 December 2023).

Ofsted. 2023. *T Level Thematic Review: Final Report*. Ofsted at https://www.gov.uk/government/publications/t-level-thematic-review-final-report/t-level-thematic-review-final-report (Accessed 8 December 2023).

Pearson. 2023. *About Us: Our Qualifications History*. https://qualifications.pearson.com/en/about-us/about-pearson/our-history.html (Accessed 29 March 2023).

Percy, E. 1945. *Higher Technological Education*. Ministry of Education.

Protect Student Choice. 2023. In https://www.protectstudentchoice.org (Accessed 4 December 2023).

Robbins Report. 1963. Higher Education. In https://education-uk.org/documents/robbins/robbins1963.html (Accessed 28 April 2024).

Sainsbury, D. 2016. *Report of the Independent Panel on Technical Education*. DFE at https://assets.publishing.service.gov.uk/government/uploads/system/uploads/attachment_data/file/536046/Report_of_the_Independent_Panel_on_Technical_Education.pdf (Accessed 4 April 2023).

Social Mobility Commission. 2020. *Apprenticeships and Social Mobility Fulfilling Potential*. Available in https://www.gov.uk/government/publications/apprenticeships-and-social-mobility-fulfilling-potential (Accessed 28 May 2024).

The Education Act. 1902. (Edw. VII, c 42) London.

———. 1944. (Georg IV, c 31) London.

UKCES. 2012. *UK Commission's Employer Skills Survey 2011: UK Results*. UKCES. Available in https://assets.publishing.service.gov.uk/media/5a7dd568e5274a5eaea669ca/ukces-employer-skills-survey-11.pdf (Accessed 28 May 2024).

Universities, UK. 2016. *Supply and Demand for Higher Level Skills, UUK*. Available in https://www.timeshighereducation.com/sites/default/files/breaking_news_files/supply_and_demand_for_higher_level_skills.pdf (Accessed 28 May 2024).

Waitt, I. 1980. *College Administration*. London: NATFHE.

Wolf, A. 2011. *Review of Vocational Education*. DFE at https://www.gov.uk/government/publications/review-of-vocational-education-the-wolf-report (Accessed 4 April 2023).

CHAPTER 3

Addressing the Challenges of Technical Education: The Potential of Further Education College Partnerships

Neil Raven

Introduction

It is widely recognised that more people across England need to be equipped with technical skills. For many commentators, this imperative is allied to the low levels of economic growth that the country has experienced for more than one and a half decades. In this respect, a recent *Financial Times* editorial (2023) described 'a major shortage' of skilled workers, whilst an investigation by the House of Commons (GOV.UK 2023b, 6) observed how 'businesses are reporting difficult[ies] recruiting employees with the relevant skills'. In particular, this investigation made reference to the Open University's 2022 Business Barometer, which 'estimated that 78 per cent of UK organisations suffered a decline in output, profitability, or growth, as a consequence of a lack of available skills' (Open University 2022). An article in the *Economist* (Anon 2022a, n.p.) made a comparable point, highlighting that between 2009 and 2019 Britain's 'productivity growth rate was the second slowest in the G7' group of nations.

N. Raven (✉)
School of Education, The University of Bristol, Bristol, UK

J. Baldwin et al. (eds.), *The Future of Technical Education*,
https://doi.org/10.1007/978-3-031-70756-8_3

Whilst there is a need to encourage the wider acquisition of technical skills at all levels, much emphasis is directed towards increasing the numbers with high-level skills, namely those above level 3 (Anon 2022a). In a presentation at a Higher Education Policy Institute (HEPI) and Advance HE symposium (10 January 2024), David Hughes, the Chief Executive of the Association of Colleges (a membership organisation that represents English further education colleges [FECs]), observed that skills needs at level 3 and above are not being met. From an international perspective, Hunt and Atfield (2019) found that compared with many other European countries, as well as Australia, New Zealand, Japan and the United States, the proportion of the population in the UK with level 4 and 5 qualifications is small. Building the case for addressing this comparative shortfall, Hunt and Atfield (2019, 88) highlight evidence that post-18 education not only 'increases the productivity of individuals' but has a positive 'spill-over effect for colleagues and others working in the wider industry or network'.

Although economic growth and productivity are prominent concerns amongst many commentators, some also consider other drivers for supporting the wider take-up of higher-level technical qualifications. Notably, those associated with enhancing social mobility and equality of opportunity. Field and Tahir (2022, n.p.) discuss an 'inequitable polarisation between those who go to university and the rest of the population'. 'These differences in education', it is argued, 'translate into huge differences in earnings'. Higher technical qualifications 'could offer a way to reduce this earnings gap'. Supporting this claim, the authors reference a recent study which found that 'level 4 and 5 qualifications can lead to earnings that compare well with degrees'.

Enabling more people across the country to access these qualifications could also address a related component of the productivity and social mobility challenge faced in England. This concerns the wide geographical differences in the take-up of HE between regions. Whilst not a perfect correlation, what are described as poorer 'left-behind' areas, tend, Pope et al. (2022, 4) observe, 'to be those with lower skill levels'. Supporting this, a study by the Resolution Foundation (2023) presented empirical evidence on the difference to an area's levels of productivity that an increase in the numbers holding higher-level qualifications can make (see also Nathwani et al. 2024), whilst a range of studies have shown a positive association between income and a higher education (The Sutton Trust

2021; Transforming Access and Student Outcomes in Higher Education 2023).

Finally, higher-level skills offer a range of non-economic benefits. Whilst discussion has tended to focus on those gaining a degree (Pope et al. 2022), a study commissioned by the Department for Education widened the remit to include post-18 education more generally, 'such as intermediate and higher-level taught courses delivered through further education, as well as vocational education and training' (Hunt and Atfield 2019, 87). 'Post-18 education', it was found, 'increases the propensity to vote', and to be politically engaged and active more generally'. It is also associated with increased 'altruistic volunteering', and a sense of well-being and a higher quality of life.

Role of Further Education Colleges

Moreover, there is wide recognition that FECs should to play a pivotal role in addressing the higher-level skills challenge the country faces. The FE sector, Lisauskaite et al. (2021, 5) note, 'is identified as a potential source of solutions' for the UK's 'labour market challenges [in] providing important vocational and technical skills needed of their workers by employers'.

However, there is some debate on the precise part that FECs should play in this context. For some, including Pheonix (2018, 42), their focus should be on delivering 'high quality provision at levels 2 and 3', and routeways that enable progression onto level 4 and above qualifications offered by universities. This contrasts with the role recognised in the Augar review (2019), which charges the FE sector with 'providing higher technical and vocational qualifications' (McVitty 2019). Arguably, it also fails to recognise FE colleges as 'anchor institutions in their local communities' (Laczik and Patel 2023, n.p.), with the potential to respond to local and regional skills needs. Supporting this assessment, Laczik and Patel (2023, n.p) reference a 2015 review of Welsh FECs that also offered HE provision. 'Further education institutions', it was observed, 'have a strong record in developing relations with local employers and developing programmes that meet higher level skill gaps', and can do so in flexible ways that 'fit around people's lives' (GOV.Wales 2015, 7).

Laczik and Patel (2023, n.p.) also draw attention to the potential of FECs to respond to the widening access agenda, as well as that associated with geographical inequalities in access to higher-level skills. 'HE in FE', it is noted, 'attracts participation from underserved and disadvantaged groups that otherwise would not engage with HE studies'. Colleges, it is added, are capable of delivering 'intermediate and higher courses flexibly and locally', which can enable them to contribute to widening access to higher education. A comparable observation is made in the review of Welsh FECs (GOV.Wales 2015, 2), whilst Parrett (2019, n.p) discusses how 'colleges that sit at the heart of their communities open up choice and opportunity for people of all ages, abilities and interests'. This, it is observed, 'goes way beyond people simply gaining qualifications. It's about transforming entire regions by offering quality educational pathways which lead to increased aspiration and social mobility'. This assessment is consistent with findings from my own studies that have highlighted the key role that FE colleges can and do play in widening HE access and progression (Raven 2023, 2021a, b).

Concept of Collaboration

However, whether FE colleges are best able to provide higher-level technical education in a competitive environment or a collaborative one is contentious, with the competition having been in the ascendence in recent decades. In describing English post-16 education and training policy, Robson et al. (2024, 6) observe 'an explicit emphasis on market logic. The government's role', it is noted, 'is seen as managing the market and maintaining market conditions through regulation with minimal centrally managed coordination'. This, it is argued, 'positions FE and HE in competition with each other as a way of responding to both consumer and economic needs'. The same would appear to apply to the conditions in which individual colleges are encouraged and expected to function, since competition is viewed as ensuring the most effective use of scarce recourses and that colleges place the student at the core of their activity (Keep 2018; GOV.UK 2016a).

Yet, a number of studies have explored the rationale and potential for local and regional collaborations of post-16 providers, although these have tended to focus on FE and HE collaboration. Laczik and Patel (2023, n.p.) argue that 'effective utilisation and coordination of the strengths of both FE and HE' can help to address national social and economic priorities, notably 'meeting skills needs'. A comparable argument underpins Pheonix's 2018 study, where the case is made that colleges should specialise on provision up to and including level 3. A subsequent study by the same author (Pheonix 2023, 6–7) advocated a different scenario but one still based on a 'joined up post-16 education system,' where the 'roles for sixth forms, colleges and universities [are] clearly defined'. Colleges, it was argued, could specialise according to 'local need', which might include 'technical provision through applied qualifications', and which could 'aligned to level 5 and 6 qualifications within local universities'. Alway et al. (2022, 21) also focus on FE and HE collaborations, although their argument acknowledges the mixed economy reality of many FE colleges, in offering both further and higher education. Here, the emphasis is on recognising that FECs are 'part of a wider [local] educational landscape', which works best when collaborating and sharing, rather than competing.

In comparison, cross-college collaborations have received little attention. This said, a study conducted by the Education and Training Foundation (Anon 2022b, see also Carroll 2022, n.p.) reported on a two-college partnership approach to supporting learners with SEND (Special Educational Needs and Disabilities). This drew attention to the benefits that collaboration between FECs could afford, including the provision of a 'joint offer', and the sharing of facilities. However, given the paucity of studies, there is a clear rationale for a closer consideration of the case for cross-college collaboration and whether such partnership working has the potential to address England's chronic skills shortage.

The Case Study

One recent initiative that aimed to address regional skills needs and support progression to higher-level training provides a suitable case study. The author was involved in conducting a summative evaluation of this large regional collaborative project which brought local colleges, as well as an employers' organisation and a number of local authorities, together in order to develop and deliver of a series of short technical courses.

The evaluation explored both the challenges to collaboration and the enablers, as well as assessing what was achieved. It also considered the lessons that this initiative offers for the wider FE sector in advancing the technical education agenda.

Context

In 2022, the UK government sought applications from collaborations of local FE providers across England for funding for the 'design, development and delivery of new skills provision' (GOV.UK 2022a, 8). The funding was available for the financial year 2022–23 and linked to the aim (expressed in the Skills for Jobs White Paper 2021) of 'reshaping the technical skills system to better support the needs of the local labour market' (GOV.UK 2022a, 5). Funding included support for procurement of equipment and infrastructure, including 'upgrading or repurposing facilities', as well as 'supporting the professional development of teaching staff' in FE colleges (GOV.UK 2022a, 12). Emphasis was placed on new provision, with the expectation that 'none' of the 'activities' concerned with the initiative would 'reflect business as usual'. Accompanying this was the requirement that new provision should benefit 'learners and employers in more deprived communities' (GOV.UK 2022a, 8).

One successful application came from the Greater Manchester Colleges Group, comprising nine FE colleges, along with the LTE (Learning, Training and Employment) Group, and the Greater Manchester Learning Provider Network (GMLPN is a network of independent training providers, FE colleges, schools and universities based in the Greater Manchester region). The bid was endorsed by the Greater Manchester Chamber of Commerce and supported by the Greater Manchester Combined Authority. The sectorial focus of this application was on green construction (GOV.UK 2022b). Its objective was to develop a series of week-long innovative courses that would support learner progression onto level 3, 4 and 5 units, or full courses at advanced and higher level. The planned subject areas concerned digital and drone surveying, retrofitting, renewable technologies, sustainable and alternative sources of energy, and building information modelling.

Method and Approach

Approach

The main element of the evaluation was qualitative in nature and sought the views of those involved in the design, development and delivery of the short courses from each of the participating colleges, as well as the Chamber of Commerce, Combined Authority and GMLPN. This was supplemented by a survey of the first cohort of students to complete the new courses, which used a mixed (qualitative and quantitative) approach.

Method

The method of data collection used in the main evaluation was the (asynchronous) email interview (Gibson 2017; Hawkins 2018; Dahlin 2021). Whilst the same set of initial questions were posed to all participants, a supplementary round of questions were also asked. These were based on the original set of replies each interviewee had provided, and were aimed at seeking clarification on key points raised, as well as providing an opportunity for participants to elaborate on their initial replies. This method of data collection was selected because it had been adopted, and found to be effective, in an evaluation of an earlier initiative involving a comparable group of stakeholders. It was also considered suitable for this group of participants since, as busy professionals, it allowed them to respond in their own time and use a medium (email) they were familiar with.

In terms of practicalities, participants were informed of the evaluation by the project lead. Once their permission to take part in the evaluation had been secured, and their agreement to be contacted by the researcher given, an email detailing the evaluation and approach that would be adopted was sent to them by the researcher. Following their reply, an email containing the initial—and main—set of questions was sent. This included a date for emailing their responses.

A questionnaire was used to gather the student perspective. This included both closed and open-ended questions. Questionnaires are a widely used method of gathering feedback from students, and one that has previously been used by the researcher (Raven 2015). In this instance, the questionnaire was sent to a contact in each of the colleges who was overseeing the development of the short course. This email included guidance on the distribution of the questionnaire. The contact was asked to send the questionnaire electronically to the short course tutor, who was asked to distribute it to all the students on the course. In turn, the students were instructed to return their completed questionnaires to the tutor. Once the tutor had received all completed questionnaires, these were to be emailed to the researcher.

Participants

In total, 13 responses were received for the main evaluation, including from eight of the participating FE colleges, as well as the Chamber of Commerce, Combined Authority and GMLPN. The feedback was sent on behalf of all of those from the same institution/organisation who had been involved in the project, rather than as a representation of a particular individual's views. In terms of the nature of answers received, the majority of participants provided detailed responses, with nine of the 13 completing the follow-up round of questions.

The questionnaire was completed by 78 students on eight of the short courses, which represented the majority of learners. All the returned questionnaires were deemed suitable for analysis. In most instances, all questions had been addressed. In the remainder of cases, most questions (baring one or two) had been answered.

The fieldwork using the questionnaire was conducted between April and early July 2023, and timed to be distributed at the end of each course. The email interviews commenced in June 2023, which coincided with the end of the funding period; although the interview process remained open until the start of September 2023, since it was recognised that some participants were likely to be on leave during parts of the summer.

Findings

Short Course Development and Delivery

The most immediate and direct achievement observed by interviewees was the development and delivery of the short courses. 'The project', one interviewee observed, 'showed that a range of short-courses could be successfully delivered by bringing employers and providers together'. A second interviewee emphasised that these were 'high quality' courses of 'industry standard'.

Interviewees also talked about high demand for the courses and the positive feedback received from those who had taken them. This assessment was confirmed in the responses provided by students in the end of course questionnaire. High levels of learner satisfaction (satisfied or very satisfied responses on a five scale rating) were returned for all surveyed courses (numbering 8), in terms of how they were taught (an average score of 4.6 out of a maximum of 5) and delivered (relating to course timing, with an average score of 4.3), and the appropriateness of their duration (length of classes, 4.4). Similarly, the content was judged to be of a high and suitable quality (in terms of the subjects covered, 4.6, the demands on their time, 4.5, and the level the course was pitch at, 4.6). A clear majority were also either satisfied or very satisfied with the support provided by their tutors (4.8), whilst 97 per cent of those who replied stated that they would recommend their course to others.

Increased Training Provision across Greater Manchester

Taking a wider perspective, interviewees talked about the project enhancing training provision in the region. 'The range of short courses now available across Greater Manchester can', one interviewer (i2) observed, 'only benefit the construction sector and upskill the workforce'. Another considered that the project had helped 'to address skills needs in a specialist area of the construction sector'. These assessments appear consistent with a response provided by students to one of the questions they were posed. More than 93 per cent of those who replied considered that the training they had received on their course had been useful. Amongst the key reasons given for this assessment concerned the knowledge gained, along with its relevance and practical application.

Evidence collected from the student questionnaires also indicated that many who enrolled on these short courses were from the target group. Of those who replied, over 38 per cent were already employed in the construction sector, with job roles ranging from company director, supervisor, surveyor, energy assessor and retrofit co-ordinator, to property landlord, development consultant, site supervisor, and trainee assessor, as well as various intermediate and advanced level apprentices (including in construction management, bricklaying and joinery), whilst one participant was a construction tutor. In addition, almost 22 per cent were taking level 3 BTECs in construction-related subjects. Moreover, amongst the principal reasons offered for enrolling on these courses were those relating to the acquisition of knowledge and skills, as well as interest, professional development and career advancement.

Progression

A number of the interviewees highlighted the progression routes that the new courses would help to open up. Our 'short course' will, one observed, 'be a feeder [on]to our current construction courses, and [offer] a means to up-skill industry professionals'. 'The new provision', another interviewee noted, should 'stimulate demand in larger level 4 [and level] 5 qualifications, which are much needed in the sector'. Underpinning these aims, most of the students surveyed confirmed that they had plans for further training. These plans included taking up in-house training opportunities and enrolling on a range of higher level programmes. Amongst the courses mentioned were an Higher National Certificate (HNC) in construction, a degree and a degree apprenticeship.

Overall Assessment

There was general agreement that the cross-college collaboration facilitated by the project had proved successful. Collaboration, one interviewee noted, has been 'strong [and] I feel [this] has benefitted lots of colleges [by] develop[ing working] relationships'. Indeed, this respondent observed that:

> At the start of the project, I felt that the most obvious outcome would be successful short courses. To a greater or lesser extent, all of the colleges achieved this. However, the collaboration, employer engagement, [and a] new approach to modular courses are, in my view, stronger and more lasting achievements, as they have the potential to improve things for the future.

Interviewees also drew attention to the links the colleges had been able to develop with other stakeholders, notably employers, 'local council[s], manufacturers, [and] other industry partners', including 'construction companies, housing associations, building material manufactures and suppliers'. Indeed, it was noted that 'a number of the colleges [had] developed new relationships with a more diverse group of employers'. Moreover, this engagement, one interviewee added, had 'improved the short course design and delivery significantly', and ensured that the resulting course aligned with 'employer and sector needs'.

Key Success Factors

Interviewees identified a range of features of the collaboration that had been instrumental in its success. Whilst a number of these have been identified as key components in effective partnership arrangements elsewhere, including collaborations in the education sector, some appear to have been less widely reported and, as previously noted, few have been reported in relation to cross-college collaborations.

Shared Goal

In explaining the project's success, one interviewee highlighted the 'desire [amongst] all partners to work together', whilst another reflected on the positive experience of engaging 'with all kinds of colleagues and stakeholders'. Similarly, a third interviewee described how the project had brought 'competitors together to achieve a [common] goal', whilst another described how the collaboration had 'foster[ed] closer working to common aims across Greater Manchester'. Consistent with this finding, Tazzyman (2023, 7) talks about 'creat[ing] a shared vision and purpose', although this was in the context of collaborations between schools, colleges and universities. Exploring partnerships in the context of education and career success Anon (2017, 4–7) also draws attention to 'shared goals or aims, visions or purposes', as well as reciprocity, in terms of the benefits each partner gains 'within a collaboration'.

Project Leaders

The part played by the project leaders was also considered pivotal by interviewees. One credited the 'collaboration' between 'the Greater Manchester Colleges and the project leads' as being the 'key to success' in the delivering a suite of short courses. In this respect, the role of the project leaders included overseeing the project bid and communications with the funders, providing guidance to individual partners, facilitating meetings, updating partners, and promoting the project to wider audiences. Similarly, in commenting on cross-sector collaborations involving technical and community colleges in the United States, Soliz et al. (2021, 692 and 704) discuss the key role played by 'individuals' who span institutional boundaries and 'are able to translate across institutional differences', and 'champion' partnerships by 'motivating others through their personal leadership skills and can-do attitudes.' Also based upon a regional partnership of schools, government agencies and higher education institutions in the USA, Anon (2017, 4–7), observes how 'individual leaders can serve as advocates' of effective colloborations.

Networking and Sharing Best Practice

Various interviewees discussed the project's success in providing an 'opportunity to network [with] colleagues' from the region's colleges, along with 'construction employers', local authorities and others. This, it was noted, enabled partners to 'share ideas and best practices.' It also facilitated the 'exposure to other professionals with different skills', and helped to identify the strengths [and] weaknesses of the [planned] short courses', as well as 'areas for [course] improvement'. Another interviewee observed how networking events organised by the project leaders provided opportunities to have 'discussions on a range of topics, from staff requirements [and] skills shortages, [to] industry developments'.

Consistent with this finding, Rudd et al. (2004, vii), discuss the value of a forum for the 'exchange of ideas and good practice' amongst schools, whilst Blackadder (2023, n.p.) highlights 'cross-partner' meetings between small HE providers in enabling partners to bring in their 'specific areas of expertise'. Elsewhere, Soliz et al. (2021, 692 and 704) reference the role of 'collaborative meetings' that 'provide a critical third space in which work across sectors' can take place, and which facilitate the building of 'trust because stakeholders communicated freely.' Indeed, consistent with this idea, one interviewee from the case study observed that the professional relationships developed during the project had provided a valuable form of 'peer support.' Another described how the 'wide' level of 'involvement' in the project across the Greater Manchester region had created 'opportunities to forge [new] professional relationships with industry leaders'. Establishing contacts with a 'more diverse group of employers' will, it was judged, 'help to influence the future of [our] short course', and provide new CPD opportunities for staff.

Evidence and Data

A number of interviewees talked about the insights that had been gained through the project on the construction sector's 'demographic and skills need[s]', both 'now and in the future', and how the project had promoted a better 'understanding of industr[ial] and technological needs' across the Greater Manchester area. This, one interviewee observed, had enabled 'new technology and practices[to be] embedded into the short course[s]'.

This finding is consistent with Anon's study (2017, 4–7) which highlights the 'effective use of evidence and data' in partnership working.

For various commentators, this evidence included that gained from project evaluations, which Tazzyman (2023, 7) notes can 'demonstrate the effectiveness of a collaborative approach.' Accordingly, the evaluation of the project was identified by various interviews as a significant source of useful insights, with one interviewee observing how the evaluation experience had provided an opportunity to reflect. This, it was added, has had 'an impact upon future decision making.' A second interviewee described finding 'the evaluation process very valuable'. In particular, reference was made to the fact that 'the feedback from the evaluations has been analysed and evaluated by an independent party to the project'. Consequently, it was added, 'this has resulted in a non-bias[ed] opinion of the project and, in turn, given [us] the opportunity to reflect on the project's journey and improve its success'.

Comparative Merits of Collaboration

A key area for discussion concerned interviewees' assessment of whether a collaborative approach to provision could better meet local and regional skills needs (and learner demand) than one based upon competition between providers. Reflecting the views of other participants, one interviewee observed how 'the project ha[d] shown that results are much greater with a collaborative approach'. A range of reasons for this view point were expressed by interviewees.

Influence and Bargaining Power

'Working collaboratively', it was observed, 'we were able to bring bigger and better employers to the table'. This, one interviewee reflected, 'had a real impact on the success of the project, meaning that it had real relevance for what employers in Greater Manchester want at this level. It allowed curriculum development to be employer-led'. This assessment is in line with Tazzyman (2023, 12) who discusses how collaborations of education providers, including colleges and universities, can 'extend the reach of individual partners.' In addition, for some interviewees this influence was also seen in terms of enhanced spending power. 'By collaborating', one participant argued, 'we had critical mass and bigger spending power to purchase better support for the project'.

Pooling Resources and Skilled Staff

Similarly, another interviewee observed how 'this initiative [offers] a good model for local areas to better pool resources'. Along similar lines, a report by the Education and Training Foundation (2022b) that considered a number of two-college partnerships aimed at supporting learners with special educational needs and disabilities, highlighted how such joint ventures had benefitted from the sharing of facilities. However, interviewees in the case study also highlighted the potential to share the expertise of skilled staff, particularly, one interviewee noted, 'in sectors where recruitment for specialist staff is especially difficult'. 'As it gets harder to get skilled staff', another interviewee observed, collaboration of the kind developed by this project is 'the only way we [can] support each other to meet the skills gap'.

Minimising Duplication and Facilitating Specialisation

Interviewees also discussed how the collaborative approach adopted by the project had helped reduce the risk of the same offer being made by multiple providers. 'Collaboration', one participant observed, 'enables parties to see what is happening, and prevents [the] duplication of activity. Working with others at a local [and] regional level is a great way to prevent several different variations of the same solution being devised'. Although concentrating on HE providers and widening participation, Tazzyman (2023, 12) also talks about collaborations reduc[ing] duplication, 'streamlining communications and delivering a more joined-up offer'. Beyond this, one interviewee highlighted how collaboration has the potential to facilitate specialisation in provision, with 'each college offering a certain area with[in] a [wider] discipline'.

Lessons for the Wider Sector

Interviewees considered that the project represented 'a good trailblazer' in cross-college and cross-organisational collaboration. 'For the sector', one interviewee observed, 'this project offers an example of an approach to the development of short courses that meets local needs, and provide[s] opportunities for progression from level 3 to level 4'. Another considered that it presented 'a model of how to begin to prepare staff to deliver higher technical qualifications through a staged process'.

More Time for Development

In addition, participants identified various lessons that the project could provide for other FECs who might be exploring cross-college collaborations. Prominent here was the need to allow more time in the development phase. Whilst the 'project worked well', one interviewee noted, there would have been value in having 'more time to discuss the short courses' being developed by 'other colleges'. A comparable point was made by another participant who called for 'more hours to be dedicated to the curriculum design and development part', since, it was observed, 'this is a very time-consuming aspect of the project and required more hours than allocated'. Any collaborative venture, it was argued, needs to recognise and take account of the time involved in developing new provision.

Time and Timelines

More generally, there was a need for more realistic timelines, including in terms of project spend. 'There was very limited time to plan and execute the spend before implementation of the programme', one interviewee observed. 'Lead times on ordering and purchasing of equipment, and relevant staff training, were not supported by the timeline presented, and this was really challenging'. One suggested response in this respect was for each college to invite other partner providers to talks by the industry specialists they had contacts with. 'May be this should be an expectation in the true spirit of collaboration in future projects', it was suggested. Consistent with this, Rudd et al. (2004, n.p.) discuss the 'practical impediments' to collaborations associated with time, whilst Tazzyman (2023, 6–7) highlight the tendency with collaborative ventures to 'underestimating the time needed' to build effective partnerships.

Project Leadership

Interviewees also talked about the need to have a 'willing project lead' who could encourage, oversee and facilitate collaboration. Similarly, emphasis was placed on the benefits of 'having employers who are part of the project' and 'who share the same interests [and] common goals'. The Education and Training Foundation (2022b, 3) also emphasise the need for colleges involved in partnership working to have 'shared values, vision and ethos'.

Early Collaborative Working and Documenting Best Practice

Less widely reported in the literature, the interviewees for this case study draw attention to the value of encouraging collaborative working from the outset of a project. 'I felt', one participant noted, 'that we worked individually for a long while to think about the offer and what that would be, and then came back together to look at it collaboratively. I think that had we worked on the development across the collaboration, rather than all individually, it would have been a good approach'. Another interviewee emphasised the need to ensure that the early development stage involves the identification and mapping of 'definitive pathways, so [that] each college offer is unique and distinct, with no or minimal overlap'. Linked to this was the benefit of ensuring that 'the best practices and successful strategies employed' from the start and throughout the project are 'documented'.

Funding

One area discussed by various commentators, as well as the interviewees in the case study, concerned the need to secure adequate project funding. For interviewees, this included being able to cover staff time, especially in the development phase. Indeed, adequate funding was judged essential in facilitating effect collaborative ventures. Similarly, Rudd et al. (2004, n.p.) talk about how insufficient levels of financial support can act as a 'barrier to cross-institutional collaborations'.

Discussion

The Potential of Collaboration

Arguably, the project's success, as recognised by interviewees and in the feedback provided by students, illustrates the potential of FE colleges to respond to the technical skills agenda. It also presents an effective approach for doing this, in terms of cross-college collaborations. Indeed, it can be suggested that this approach has the capacity to harness the traditional strengths of FE colleges, with strong ties to their local communities, a detailed understanding of the barriers faced by a significant number of people in accessing and taking advantage of post-compulsory education, and an expertise in responding to learner needs, as well as having established working relationships with local employers. In addition, the approach has the potential to address some of the challenges that colleges can encounter when working on their own. Collaborations can support engagement with a wider range of employers, and provide access to a greater number of skilled staff, as well as enable partner institutions to take advantage of economies of scale and exercise greater bargaining power in negotiating with suppliers, both of which are especially relevant in technical education where costs per student tend to be comparatively high (GOV.UK 2016b).

The Case Study Finding in Context

Yet, when interpreting these findings it is important to recognise the modest nature of the project, in terms of its focus on one sector and, in particular, one component of that sector, as well as the scale of the new provision generated. However, as various interviewees observed, the project was able to demonstrate 'proof of concept', and afford a 'blue print' for others across the FE sector to draw upon. Moreover, interviewees expressed enthusiasm for adopting the same model in addressing other areas where higher-level technical skills are needed, locally and regionally. Although this was couched in the context of long-standing challenges faced by FE colleges across England, notably insufficient funding. Indeed, some interviewees argued that the project had highlighted the need for the sector to be adequately financed. 'Sufficient funding', it was observed, is required to enable 'the hiring and retention of high-quality teaching staff, and investment in infrastructure, equipment and student support'.

Policy Direction

Recent policy developments in the form of Local Skills Improvement Plans (LSIPs) may afford some optimism in this respect, and may also suggest support for the type of collaborative approach developed in this project. Whilst much government rhetoric has focused on market forces and competition between providers, LSIPs place greater emphasis on colleges working together to address local skills deficits (GOV.UK 2023a). Indeed, a number of the interviewees talked about the approach adopted by project partners being consistent with that expected in LSIPs.

Introduced in 2023, and based on 38 areas across England, LSIPs involve organisations representing employers coming together to identify local skills needs. Local FE colleges, as well as other training providers, are then required to respond by providing training opportunities able to address these needs. To support the collective college response in each area, a Local Skills Improvement Fund has been made available. This, it is stated, can be used to 'investment in new facilities and equipment', and support the 'development' and teaching 'of new courses, particularly [those offered] at levels 3 to 5' (GOV.UK 2023a, n.p.).

Case for Caution

However, whilst LSIPs have generally been welcomed by the FE sector, there remains a need for caution. Concerns have been raised about the employer directed nature of this initiative. Even the Confederation of British Industry (which is an employer-based pressure group) observes that 'asking businesses what they need does not always translate easily into a learning and development programme' (Anon 2022c, 4). Moreover, little reference is made to the role, or provision, of careers guidance in supporting those who could benefit from the new skills that training opportunities faciliated by LSIPs may provide (Keep 2023). Beyond this, it can be suggested that the skills that businesses consider to be needed now may not be the same as those required in the future and once students have completed a full HE level course.

Doubts have also been raised about the level of funding being sufficient to cover the costs of developing and delivering provision capable of addressing what are likely to be a range of skills needs in a number of different sectors. Moreover, the funding is confined to a two-year period (GOV.UK 2023b). Although LSIPs recognise the key role that FE colleges can play in addressing the skills deficit, and enhancing access to higher levels of education and training, including amongst those from areas of deprivation, this latest initiative is likely to fall short of what is really required to make a lasting impact on a deep seated problem.

What Is Needed

This said, the case study has highlighted both the potential of FE colleges to play a key role in addressing this challenge and what is needed to sustain it. The policy prescription for a coherent technical education landscape seems clear: sustained and adequate funding that enables and encourages collaboration. Whether it is forthcoming is, as ever with the English FE sector, a rather different matter.

References

Alway, P., L. Cooper, N. Day, L. Morgan. 2022. *Going Further and Higher: How Collaboration Between Colleges and Universities Can Transform Lives and Places.* Sheffield Hallam University and the Civic University Network. https://governance4fe.co.uk/going-further-and-higher-how-collaboration-between-colleges-and-universities-can-transform-lives-and-places/ (Accessed 21 May 2024).

Anon. 2017. *A Compilation of Research on Cross-sector Education and Career Partnerships, The Regional Educational Laboratory West, San Francisco.* https://ies.ed.gov/ncee/edlabs/regions/west/relwestFiles/pdf/APECS_5-2-1_Literature_Compilation_clean.pdf.

———. 2022a. Britain's Productivity Problem Is Long-standing and Getting Wose. *The Economist* (June 9). https://www.economist.com/britain/2022/06/09/britains-productivity-problem-is-long-standing-and-getting-worse (Accessed 21 May 2024).

———. 2022b. Learners with SEND benefit when specialist and general colleges work together, new ETF report finds. *Education and Training Foundation* (June 22), https://www.et-foundation.co.uk/news/learners-with-send-benefit-whenspecialist-and-general-colleges-work-together-new-etf-report-finds/ (Accessed 19 September 2024).

———. 2022c. Skills Bill Becomes Law. *FE News* (28 April). https://www.fenews.co.uk/skills/skills-bill-clears-way-for-ifate-to-unify-system/ (Accessed 21 May 2024).

Blackadder, L. 2023. *Many Hands Project: A Collaborative Approach By Small Setting, Independent, Higher Education Providers to Student Mental Health Solutions.* Office for Students. https://www.officeforstudents.org.uk/advice-and-guidance/promoting-equal-opportunities/effective-practice/many-hands-project-a-collaborative-approach-to-student-mental-health-solutions/ (Accessed 21 May 2024).

Carroll, T. 2022. Better together: why partnerships can deliver for learners with SEND and those that teach them. *FE News* (2 July), https://www.fenews.co.uk/fe-voices/better-together-why-partnerships-can-deliver-for-learners-with-sendand-those-that-teach-them/ (Accessed 19 September 2024).

Confederation of British Industry. 2022. *Local Skills Improvement Plans. CBI Feedback to the Department for Education* (June 2022). https://www.cbi.org.uk/media/p42nbs4t/cbi-lsip-report-june-2022.pdf (Accessed 21 May 2024).

Dahlin, E. 2021. Email Interviews: A Guide to Research Design and Implementation. *International Journal of Qualitative Methods, 20.* https://journals.sagepub.com/doi/10.1177/16094069211025453 (Accessed 21 May 2024).

Editorial Board. 2023. Tackling Britain's Dearth of Skills and Workers. Fixing Education, Training and High Inactivity Is Key to Boosting UK Productivity. *Financial Times* (28 February), https://www.ft.com/content/e790c414-3640-4285-a3c3-ce0a222068ae (Accessed 21 May 2024).

Field, S., and I. Tahir. 2022. *The Missing Middle of Higher Technical Learners.* Institute of Fiscal Studies (22 September), https://ifs.org.uk/articles/missing-middle-higher-technical-learners (Accessed 21 May 2024).

Gibson, L. 2017. 'Type Me Your Answer.' Generating Interview Data Via Email. In *Collecting Qualitative Data: A Practical Guide to Textual, Media and Virtual Techniques*, ed. V. Braun, V. Clarke, and D. Gray, 211–212. Cambridge University Press.

GOV.UK. 2016a. *Understanding the Further Education Market in England.* BIS Research Paper 296. Department for Business, Innovation and Skills. https://assets.publishing.service.gov.uk/media/5a74d9a5e5274a3cb2867af7/bis-16-360-fe-market-england.pdf (Accessed 21 May 2024).

———. 2016b. *Report of the Independent Panel of Technical Education.* https://assets.publishing.service.gov.uk/media/5a8075a7e5274a2e8ab50444/Report_of_the_Independent_Panel_on_Technical_Education.pdf (Accessed 21 May 2024).

———. 2022a. *Strategic Development Fund. Application Guidance*, Department for Education. https://assets.publishing.service.gov.uk/media/63107f41d3bf7f739da926da/WITHDRAWN_-Strategic_development_fund_2022_prospectus.pdf (Accessed 21 May 2024).

———. 2022b. *Strategic Development Fund: 2022 to 2023 Financial Year.* Department for Education. https://www.gov.uk/government/publications/strategic-development-fund-awards-2022-to-2023/strategic-development-fund-2022-to-2023-financial-year (Accessed 21 May 2024).

———. 2023a. *Skills and Labour Shortages.* House of Commons Library. https://researchbriefings.files.parliament.uk/documents/CDP-2023-0001/CDP-2023-0001.pdf (Accessed 21 May 2024).

———. 2023b. *Local Skills Improvement Plans. Stage 2 Guidance: Objectives.* Department of Education. https://assets.publishing.service.gov.uk/media/654b7980e70413000ffc49fb/Local_skills_improvement_plans_-_stage_2.pdf (Accessed 21 May 2024).

GOV.Wales. 2015. *Review of Higher Education in Further Education Institutions*, Welsh Government, Reseaerch Document. https://www.gov.wales/sites/default/files/publications/2018-02/review-of-higher-education-in-further-education-institutions.pdf (Accessed 21 May 2024).

Hawkins, J.E. 2018. The Practical Utility and Suitability of Email Interviews in Qualitative Research. *Nursing Faculty Publications*, 24. https://core.ac.uk/download/pdf/217291439.pdf (Accessed 21 May 2024).

Higher Education Policy Institute and Advance HE. n.d. *The Future of Post-18 Education and Whether It Should Be 'Tertiary'*, Webinar, 10 January 2024.

Hunt, W., and G. Atfield. 2019. *The Wider (Non-market) Benefits of Post-18 Education for Individuals and Society*. Department for Education, https://assets.publishing.service.gov.uk/media/5f36b9bed3bf7f1b1c0d836d/Benefits_of_post_18_education.pdf (Accessed 21 May 2024).

Keep, E. 2018. Market or System—What's the Best Model for FE?, *FE Week* (16 April). https://feweek.co.uk/market-or-system-whats-the-best-model-for-fe/ (Accessed 21 May 2024).

Keep, E. 2023. Getting LSIPs Right. Why Adjusting the Current Model Could Reap Rewards for Skills in England. *FE Week* (7 July). https://feweek.co.uk/getting-lsips-right-why-adjusting-the-current-model-could-reap-rewards-for-skills-in-england/ (Accessed 21 May 2024).

Laczik, A., and J. Patel. 2023. The Evolving Tertiary Space in the UK: Meeting the Skills Needs Through HE in FE. Higher Education Policy Institute Blog (9 August), https://www.hepi.ac.uk/2023/08/09/the-evolving-tertiary-space-in-the-uk-meeting-the-skills-needs-through-he-in-fe/ (Accessed 21 May 2024).

Lisauskaite, E., S. McIntosh, S. Speckesser, and H. Espinoza. 2021. *Going Further. Further Education, Disadvantage and Social Mobility*. The Sutton Trust and Centre for Vocational Education Research. https://www.suttontrust.com/wp-content/uploads/2021/10/Going-Further.pdf (Accessed 21 May 2024).

McVitty, D. 2019. *The Augar Review: The Essential Overview for HE* (30 May), *WonkHE*. https://wonkhe.com/blogs/the-augar-review-the-essential-overview-for-he/ (Accessed 21 May 2024).

Nathwani, T., J. Bermingham, S. Dent, and S. Donnelly. 2024. Inclusive Growth: Can Data Support Policymakers and the Higher Education Sector Meet This Objective?', blog, *Higher Education Statistics Agency* (27 February). https://www.hesa.ac.uk/insight/27-02-2024/inclusive-growth-data-higher-education-sector (Accessed 21 May 2024).

Parrett, S. 2019. Why Collaboration Is Key to Colleges Success in 2020. *TES Magazine* (18 December). https://www.tes.com/magazine/archive/why-collaboration-key-colleges-success-2020 (Accessed 21 May 2024).

Pheonix, D. 2018. *Filling in the Biggest Skills Gap: Increasing Learning at Levels 4 and 5*. Higher Education Policy Institute Report 110. https://www.hepi.ac.uk/wp-content/uploads/2018/08/EMBARGOED-UNTIL-23-AUGUST-2018-Filling-in-the-biggest-skills-gap-Report-110_Final-web.pdf

Phoenix, D. 2023. *Connecting the Dots: The Need for an Effective Skills System in England, Higher Education Policy Institute. Report 167*. https://www.hepi.ac.uk/2023/11/23/new-report-calls-for-a-skills-strategy-to-coordinate-post-16-provision-and-support-collaboration-between-education-providers/ (Accessed 21 May 2024).

Pope, T., E. Shearer, and P. Hourston. 2022. *Levelling Up and Skills Policy. How Qualifications and Training Can Help Boost Regional Productivity*. Institute for Government. https://www.instituteforgovernment.org.uk/publication/levelling-up-skills-policy (Accessed 21 May 2024).

Raven, N. 2015. The Impact of Mentoring Reconsidered: An Exploration of the Benefits for Student Mentors. *Research in Post-Compulsory Education* 20 (3): 280–295.

———. 2021a. Widening HE Access from FE Colleges: The Key Role Played By Subject Tutors. *Perspectives: Policy and Practice in Higher Education*. https://doi.org/10.1080/13603108.2021.1961173 (Accessed 21 May 2024).

———. 2021b. Making a Difference: Insights into Effective HE Progression Practices in Further Education Colleges. *Widening Participation and Lifelong Learning* 23 (1): 79–101.

———. 2023. Supporting HE Progression from FE Colleges: Learning from the Student Voice. In *One Voice: Fusing Diverse Perspectives for Collective Action*, ed. N. Raven, 55–77. Peterborough: Forum for Access and Continuing Education.

Resolution Foundation. 2023. *Ending Stagnation: A New Economic Strategy for Britain*. Resolution Foundation and Centre for Economic Performance, London School of Economics. https://economy2030.resolutionfoundation.org/wp-content/uploads/2023/12/Ending-stagnation-final-report.pdf (Accessed 21 May 2024).

Robson, J., L. Sibieta, S. Khandekar, M. Neagu, D. Robinson, and S. J. Relly. 2024. *Comparing Policies, Participation and Inequalities Across UK Post-16 Education and Training Landscapes*. Nuffield Foundation. https://epi.org.uk/publications-and-research/uk-nations-education-and-training/ (Accessed 21 May 2024).

Rudd, P., A. Lines, S. Schagen, R. Smith, and A. Reakes. 2004. *Partnership Approaches to Sharing Best Practice*. National Foundation for Educational Research. https://www.nfer.ac.uk/publications/partnership-approaches-to-sharing-best-practice/ (Accessed 21 May 2024).

Soliz, A., C. DeLoach, and H. Mesa. 2021. How Do Community and Technical Colleges Build Cross-sector Collaborations? *The Journal of Higher Education* 94 (6), 691–719. https://www.tandfonline.com/doi/full/10.1080/00221546.2023.2171212 (Accessed 21 May 2024).

Tazzyman, S. 2023. *The Benefits and Barriers to Collaborative Access Activity By Higher Education Providers. A Report for the OfS*. https://www.officeforstudents.org.uk/media/3ad6de14-b501-4b3e-be63-d77e23e9fffa/benefits-and-barriers-to-collaborative-access-activity-by-higher-education-providers.pdf (Accessed 21 May 2024)

The Open University. 2022. *Business Barometer. Navigating the Skills Landscape*. The Open University and British Chambers of Commerce. https://british-chambers.org.uk/wp-content/uploads/2023/09/The-Open-University-Business-Barometer-2022-report.pdf (Accessed 21 May 2024).

The Sutton Trust. 2021. *University and Social Mobility: Summary Report*. The Sutton Trust. https://www.suttontrust.com/wp-content/uploads/2021/11/Universities-and-social-mobility-final-summary.pdf (Accessed 21 May 2024).

Transforming Access and Student Outcomes in Higher Education (TASO). 2023. *The Value of Higher Education. Rapid Evidence Review and Initial Data Analysis*, Phase 1 Report. https://taso.org.uk/wp-content/uploads/Taso-full-report-the-value-of-higher-education.pdf (Accessed 21 May 2024).

CHAPTER 4

Credit Where Credit's Due: Ascribing Credit to Apprenticeships—Facilitating Progression to Higher Education

Darryll Bravenboer, Mandy Crawford-Lee, and Clare Dunn

Introduction

Apprenticeships in England, while defined by qualification level, are not defined by total learning time, notional learning hours or guided learning hours (GLH) as with other regulated qualifications and accredited programmes. Given it is more difficult to determine the total number of hours

D. Bravenboer
Middlesex University, London, UK
e-mail: d.bravenboer@mdx.ac.uk

M. Crawford-Lee (✉)
University Vocational Awards Council, Bolton, UK
e-mail: mc15ext@bolton.ac.uk

C. Dunn
The Open University, Milton Keynes, UK
e-mail: clare.dunn@open.ac.uk

J. Baldwin et al. (eds.), *The Future of Technical Education*,
https://doi.org/10.1007/978-3-031-70756-8_4

needed to complete an apprenticeship (and not just for any mandated qualification required), and therefore ascribe credit, this chapter proposes a mechanism and approach that will recognise the full extent of the amount of learning needed both on- and off the job to complete an apprenticeship and remove some of the existing barriers to accessing HE. In addition, the differing concepts of both on-the-job and off-the-job learning, and how integrating both in apprenticeship delivery is needed, are explored. This chapter also reiterates the importance and significance of undertaking on-the-job learning/training that is workplace contextualised, in recognition of people's potential that is not bound by traditional approaches to university admission.

Indeed, an understanding of the type and value of the learning in an apprenticeship is as important as estimating the number of credits, whatever the job role, occupation, profession or industry sector. By capturing and recognising *all* the learning undertaken during an apprenticeship, and ascribing a credit value, programme admissions teams can better understand the amount of learning achieved as preparation for higher-level study. Apprenticeships when done well enable people to develop the kinds of professional and occupational learning that is perhaps at least as valuable, if not *more* valuable, than learning gained from other experiences and qualifications. As such, apprenticeship achievement (at all levels) should give parity to those who might be seeking to progress to HE programmes from the esoterically called 'non-traditional' routes. This chapter looks to give university and HE providers, admission tutors and curriculum leaders the confidence that learners will be successful and appropriately prepared and will achieve their HE qualifications as apprenticeships offer, where applicable, alignment with entry requirements to HE. Without this kind of alignment ensuring good preparation for the next level of learning, it negatively affects university admissions decisions.

Background

The advent of the twenty-first century saw a significant shift in the development and policy design of apprenticeships in England (Crawford-Lee 2022, 23). During the twentieth century, apprenticeships evolved as a 'parallel' delivery model based on day release for off-the-job training. This was a model also commonly used for most professional training until the mid-century when it moved to full-time higher education study programme followed by work-based training. However, apprenticeships became, increasingly, after a revival in the 1980s, a programme to address young people's unemployment, limited to lower-level skills, entry job roles and qualifications at Levels 2 and 3. As a result, it reinforced the notion that apprenticeships were a further education (FE) programme, the domain of traditional manufacturing and engineering industries, craft and trades, and not an appropriate route to enter professional job roles (Lester 2020).

What followed in the 2000s was a period of renewed investment and interest in apprenticeships, and the programme was relaunched in 2008 under a newly created agency of government called the National Apprenticeship Service (NAS). Given the emphasis on higher technical skills in both the Dearing (Department for Education and Employment [DfEE] 1997) and Leitch (2006) reviews, apprenticeships were extended to include higher apprenticeship 'frameworks' at skill and qualification Levels 4 and 5 but which simply followed the pre-existing delivery model of achievement of separate 'knowledge' and 'competence' qualifications—the criteria for Levels 2 and 3 apprenticeships—which meant very little engagement from higher education providers. Limits to HE involvement extended from this dual qualification requirement in addition to a lack of understanding that 'competence' could be incorporated into university qualifications, the inclusion of only 'small' qualifications in terms of credit size and low levels of funding (Anderson et al. 2012; Bravenboer 2016). It was not until the revision to the Specification of Apprenticeship Standards in England (SASE; DBIS 2013) that significant changes happened. SASE allowed for the inclusion of a single integrated qualification that assessed both 'knowledge' and 'competence' in an apprenticeship framework (Bravenboer and Lester 2016, 16), further extended the range of higher apprenticeships to Levels 6 and 7 and introduced a requirement for higher apprenticeships to include qualifications closer in size to the standards and norms in higher education. An important addition too was the inclusion in higher-level apprenticeships of professional recognition, where applicable. This new generation of higher apprenticeships would prove to be more attractive to higher education providers in terms of delivery, plus professional, statutory and regulatory bodies (PSRBs) and employers looking to create routes into the professions and upskill and reskill their own staff (PARN 2015).

An independent review of what an apprenticeship should be in England, the Richard Review (2012), led to more changes to the specification. These included employer-led 'standards-based' apprenticeships leading to a synoptic end point assessment replacing frameworks and a call for apprenticeships to become an effective progression route into professional occupations and senior-level, higher-paid work roles which, in effect, led to the introduction and launch of degree apprenticeships in 2015. HE by definition with the inclusion of a university degree at bachelor's or master's level, degree apprenticeships either retained a separate end point assessment or 'integrated' the degree to be coterminous with end point assessment and therefore did not require a separate assessment of occupational competence (Department for Business, Innovation and Skills [DBIS] 2015, 13). *The Future of Apprenticeships* (DBIS 2015), which superseded the 2013 SASE, carried over from it the extension to apprenticeships at Levels 6 and 7 along with a mandate to include trade or professional registration and recognition where applicable but did not carry over the specification for achievement of a minimum number of credits. Instead, the size of an apprenticeship became defined by its duration (not less than 12 months) and a minimum requirement for 20% off-the-job training which was equivalent to the day release model which characterised an apprentice's entitlement more than half a century before.

While many of these reforms, including the advent of an apprenticeships levy (Finance Act 2016) (first announced in 2015 and then came into being in 2017) and the setting up of the Institute for Apprenticeships (and later Technical Education), an employer-led crown non-departmental public body (NDPB), came about in the context of fiscal rather than education policy, together they communicated a change in the engagement of higher education providers in apprenticeship delivery. Combined, they would become regarded as 'game-changers' that would sizeably alter the market for higher technical and professional education and training (Anderson and Crawford-Lee 2016, 346; Crawford-Lee 2022, 45).

Introduction

It is true that the relationship between apprenticeships and HE in England has significantly fluctuated over the last two decades or more. The National Apprenticeship Service (NAS) prospectus for the Higher Apprenticeship Fund (NAS 2011), launched by the Westminster coalition government, describes universities as a source of innovation for models of delivery but also states:

> The idea of an Apprenticeship as being a route to professional status goes against the grain of universities being the "gateway to the professions". (NAS 2011, 11)

At the time, the regulatory framework that underpinned the structure of apprenticeships also bore very little resemblance to the expectations for HE programmes. The Quality Assurance Agency for Higher Education's (QAA) Framework for Higher Education Qualifications (QAA 2008) and Higher Education Credit Framework for England (QAA 2008) established common expectations regarding both the level and the credit value of HE qualifications. For example, common expectations were set that certificate of higher education qualifications amount to 120 credits at Level 4, that diploma of higher education qualifications amount to 240 credits at Levels 4 and 5, bachelor's degrees with honours amount to 360 credits at Levels 4, 5 and 6 and master's degree qualifications amount to 180 credits at level 7. However, as discussed by Anderson et al. (2012), SASE (DBIS 2011) indicated that higher apprenticeships could only be delivered at Levels 4 and 5 and the qualifications required could range widely in size from as little as 10 credits but up to 240 credits at Levels 4 and 5.

It was the revised SASE (DBIS 2013) which was specifically designed to better align the credit value of apprenticeships with HE qualifications. This included enabling apprenticeships to be approved at Levels 6 and 7 for the first time and required higher apprenticeships at Levels 6 and 7 to amount to at least 120 credits and at least 90 credits at Levels 4 and 5. In addition, SASE 2013 required apprenticeships to lead to professional recognition (where available—although it is relevant that some professions had already established 'apprenticeship-type' routes to qualifying that included higher education qualifications but with little integration with the workplace), which, in addition to the requirement to 'level' all apprenticeships, was a feature retained in the revised model of apprenticeship standards introduced subsequent to the Richard Review (2012). On the one hand, SASE 2013 allowed for the inclusion of a single integrated qualification that was not reliant on assessing 'knowledge' and 'competence' separately (Bravenboer and Lester 2016, 16); on the other, the alignment with higher education or any other qualifications regarding credit value was not retained following the introduction of apprenticeship standards in England in 2013.

The success of integrated degree apprenticeships since their introduction in 2015—hailed as "the greatest opportunity ever seen for anyone concerned with skills and employment" (Jeffrey 2016, 1)—and the recent move by the Institute for Apprenticeships and Technical Education (IfATE) towards a fully integrated model (IfATE 2022) hold the prospect of a comprehensive alignment between apprenticeships and HE for the benefit of apprentices, employers and providers alike. Furthermore, where once there was a government failure to recognise that development of skills via many of those 'traditional apprenticeships' did not merely stop at Level 3, whether in real crafts and trades or in what are now considered to be formalised professions, there is now greater acceptance that a revival of professional entry routes through apprenticeships are an important part of the discourse on education and skills policy because they have a meaningful impact on what is and what is not viable in the qualifications system in England (Crawford-Lee 2022, 4).

From the outset, degree apprenticeships were intended to bring the workplace and the HE sector together under two main policy objectives: to attract a new cohort of skilled workers equipped to grow the economy by increasing business productivity and to raise levels of social mobility in terms of supporting greater access to HE and the professions (Crawford-Lee 2022, 55). However, there is a crucial missing component in HE that is undermining the consistent recognition of apprenticeships that manifests as a barrier to career progression for individuals and inhibits meeting employer workforce needs.

The positive alignment afforded by an integrated model is arguably made possible by the fact that all apprenticeships in England are approved by 'level' as a measure of the relative complexity, challenge and autonomy required to establish occupational and professional competence. More than this, the conception of 'level' employed is the same as that which is established and used to determine the 'level' of HE qualifications and the award of academic credit in England (QAA 2021). The use of a common conception of levels across apprenticeships, HE (and other) qualifications and the professions holds the greatest prospect of enabling consistency of recognition, by employers, higher education providers and PSRBs, which in turn has the potential to remove barriers to flexible career progression.

The 'Independent Panel Report to the Review of Post-18 Education and Funding' (the Augar Report) (Department for Education [DfE] 2019) made key recommendations regarding how academic credit could support the enhanced flexibility of higher-level learning: "Learners should be able to access student finance for tuition fee and maintenance support for modules of credit-based Level 4, 5 and 6 qualifications" (p. 40). This was followed up in the 'Skills for Jobs' White Paper (DfE 2021), which first proposed the introduction of the Lifelong Loan/Learning Entitlement (LLE), which became a flagship measure in the 'Skills and Post-16 Education Act' (DfE 2022a).

> To enable greater flexibility, we must ensure that high-quality provision is normally credit-bearing thereby enabling learners to accumulate and transfer credit where appropriate between institutions and to build up to meaningful qualifications over time fitting in with their personal and work circumstances. (DfE 2021, 41)

However, at present, the identified benefits of flexibility, as a consequence of ensuring that provision is credit-bearing, are not required for apprenticeships in England. While all apprenticeship standards in England have an ascribed level, they do not require a credit value (or number of credits) to be ascribed. This means that while, on the one hand, all apprenticeships in England can be afforded consistent recognition with regards to the level of complexity, challenge and autonomy required, there is no consistent measure of the amount of learning/training required to achieve occupational or professional competence. There are established expectations that specific qualifications normally amount to a specific amount of credit at identified levels but there is no equivalent established expectation regarding the credit value, or number of credits, for the apprenticeship itself.

This matters because it inhibits flexible career progression for individuals but also because it results in national, regional and employment sector skill gaps to the detriment of UK plc. For example, Britain's National Health Service (NHS) Long Term Plan (2019) states:

> 14,000 applicants to nursing were not accepted onto courses. At a time of staff shortage across the NHS...many thousands of highly motivated and well-qualified applicants...are being turned away. (NHS 2019, 80)

It is of course not the case that all of the 14,000 applications identified by the NHS were turned away because their apprenticeship achievements were not appropriately or consistently recognised. The plan also identifies a wider issue regarding the arguably inflated tariff requirements for entry of some higher education providers. However, this chapter will argue that ascribing credit, and specifically a credit value, to apprenticeships will support more consistent recognition by higher education providers, and this will help to remove at least one of the barriers to flexible and successful career progression.

There may also be the opportunity to align the UK Government's LLE policy and apprenticeship policy in promoting the wider skills agenda. The LLE will use credit as the metric for establishing a common value for the amount of funded HE learning that is available. Individuals will have access to 480 credits (at qualification levels 4–6), which, based on current funding amounts, equates to £37,000 or circa £77 per credit. The introduction of use of credit as the measurable learning quantum for HE is predicated on the flexibilities it will afford individuals with regards to progressing their careers throughout their lives. Individuals will be able to flexibly build skills by accessing education and training delivered by a wider range of organisations in contexts that reflect the pattern of their lives. Such flexibilities will be further enhanced if the metric that is used to measure the learning quantum for HE is the same as that which is used for higher and degree apprenticeships. This will enable individuals who have completed a higher or degree apprenticeship to gain recognition more consistently for the amount of learning achieved to support lifelong learning, career development and both professional and personal aspirations.

Recognising Talent

The research that underpins this chapter was carried out in the context of work funded by Health Education England (HEE), now NHS England, an NDPB and part of the NHS, entitled 'Recognising Talent for Care: Pathways for Progression to Higher Education'. The primary aim was to establish a national industry standard for the recognition of prior learning in the health and adult social care sectors to support individuals in support worker roles access and gain entry to HE healthcare professional programmes and progress their careers. It involved working with key organisations and stakeholders to establish a commitment to promote progression to relevant HE programmes and focused on establishing a national standard for accrediting prior and experiential learning (APEL), including learning achieved through work.

As such, the work focused on establishing (as a minimum) parity of recognition for individuals seeking access to HE and progression to registered professions, within the health and adult social care sector, through non-traditional routes. In particular, there was a focus on addressing the lack of recognition of the Level 3, Senior Healthcare Support Worker apprenticeship by higher education providers. The Talent for Care (Health Education England [HEE] 2014) strategy, which identified the need to better career progression support for the healthcare support workforce, identified the following:

> The challenge is inconsistency...even those who have good access to training and education may find their qualifications or past training does not transfer from one organisation to another or enable progression into higher education. (HEE 2014, 11)

The research sought to address this inconsistent recognition by establishing a common understanding of the value of the Senior Healthcare Support Worker apprenticeship as preparation for higher-level study through the process for ascribing credit. This particular apprenticeship prepares people to work in a range of health and adult social care settings by working side by side with registered healthcare professionals on a daily basis. Learners completing this apprenticeship gain unique insight into what is required to become a registered healthcare professional, which arguably provides a far better preparation for HE healthcare programmes than more traditional entry qualifications. This is hugely significant as it speaks to the crucial missing component in HE, which is that the apprenticeship is not sufficiently recognised through traditional admissions processes. The research undertaken sought to address this issue and the barriers to progression it creates.

Although the apprenticeship was already approved by the IfATE at Level 3, further analysis was undertaken to demonstrate how the required areas of knowledge, skills and behaviours (KSBs) were aligned with the learning expectation described within the SEEC Credit Level Descriptors (SEEC 2021). The SEEC Descriptors, which are highlighted within the QAA Higher Education Credit Framework for England (QAA 2021), describe learning expectations in a coherent and consistent form (from qualification level 3 to 8) that is used by HE providers in the development of their programmes. At the same time, the SEEC Descriptors are also now specifically designed to reflect the KSB requirements of apprenticeship programme development. As a consequence, the SEEC Descriptors provide an excellent means to establish the kind of learning required by the senior healthcare support worker apprenticeship in a language that is familiar and recognised by HE providers. The outcome of this analysis was that the senior healthcare support worker apprenticeship strongly aligned with the expectations for Level 3 and particularly with areas that are significant as a preparation for progression to HE. This included problem-solving; research and enquiry; project design and development skills; ethical awareness and application; and personal responsibility and leadership.

Having confirmed the value of the senior healthcare support worker apprenticeship regarding academic level and as an excellent preparation for HE, the remaining task was to establish the number of credits that should be ascribed to the apprenticeship. The process developed by the research to ascribe the credit value (or number of credits) to the senior healthcare support worker apprenticeship, articulated below, not only seeks to provide a reasonable means to establish the credit value, or number of credits, for the senior healthcare support worker apprenticeship but is also offered as a standardised process that could be applied to any apprenticeship, at any level, in England.

The Challenge of Ascribing Credit Value to Apprenticeships

The specific regulatory environment within which apprenticeships in England are required to operate provides significant challenges regarding the development of a consistent process for ascribing credit value. These challenges are concerned with the variety of ways in which the number of required learning/training hours are described. Terms such as 'learning hours', 'notional learning hours' and 'training hours' are deemed to be synonymous as, in one way or another, they are all concerned with a measure of the time taken to engage with a learning/training activity of some kind. Similarly, no distinction is made between 'learning' and 'training' as, in the context of apprenticeships, both are intended to bring about the development of identified outcomes in the form of KSBs required for occupational/professional competence.

Any process for ascribing credit to any learning programme, module or unit, including apprenticeships, requires two key factors to be established. Firstly, the level at which learning is intended to take place as "an indicator of the relative complexity, demand and/or depth of learning" (QAA 2021) and, secondly, the credit value as a measure of the following:

> The number of credits, at a particular level, assigned to a body of learning. The number of credits is based on the estimated learning hours (where one credit typically represents 10 notional hours of learning). (QAA 2021, 1)

This means that to determine a credit value for an apprenticeship, in other words the number of credits that should be ascribed to it, there is a need to establish how many learning/training hours are required for an apprentice to achieve the KSBs (the learning outcomes) specified in the relevant apprenticeship standard. Approved apprenticeship standards in England are published on the IfATE's website (www.instituteforapprenticeships.org). Whereas the vast majority of accredited qualifications, in the UK and internationally, do specify the learning hours required to complete an associated programme of study, as already indicated, apprenticeship standards in England do not. While all apprenticeship standards specify a level (which equates to academic credit levels) and typical duration expressed in months, the number of learning/training hours required to complete an apprenticeship is not specified. While the 'typical duration' of an apprenticeship is a measure of time, usually specified in months, it does not provide a valid measure of learning/training hours required that can be used to calculate credit value. Duration is not a valid measure primarily, as it does not indicate the intensity and/or the amount of learning/training that may be undertaken over the period indicated.

In England, all apprenticeships require a combination of on- and off-the-job learning/training to enable apprentices to develop the required KSBs to achieve occupational/professional competence. For example, in the formal response to their consultation on degree apprenticeships, IfATE have stated:

> Apprenticeships work best when learning and experience in the workplace and off-the-job training delivered by the training provider are fully integrated and mutually reinforcing. (IfATE 2022)

Furthermore, the integration of on- and off-the-job learning/training is "intended to ensure that on-the-job and off-the-job training are delivered in a co-ordinated and mutually reinforcing way" (IfATE 2022).

The Education and Skills Funding Agency (ESFA) (an executive agency of the government for funding education and skills in England) Apprenticeship Funding Rules (DfE 2023) specify that to be eligible apprentices must have formal off-the-job training (which is a statutory requirement for an English apprenticeship and the responsibility of the provider) and

> The opportunity to apply these new skills in a real work environment, in a productive job role, (which is the responsibility of the apprentice's employer). (DfE 2023, 28)

This would seem to suggest that on-the-job learning/training can contribute to the development (through application) of 'new skills' (the KSBs). However, rather confusingly, the funding rules distinguish between off-the-job and on-the-job learning/training hours by stating that whereas off-the-job training

> is training which is received by the apprentice within their practical period, during the apprentice's normal working hours, for the purpose of achieving the knowledge, skills and behaviours of the apprenticeship they are undertaking.
>
> It is not on-the-job training, which is training received by the apprentice for the sole purpose of enabling the apprentice to perform the work for which they have been employed. By this we mean training that does not specifically link to the knowledge, skills and behaviours set out in the apprenticeship. (DfE 2023, 38)

This could be interpreted to mean that off-the-job learning/training must contribute to the development of the KSBs but on-the-job learning/training (through application in the workplace) may contribute to the development of the KSBs *or* may not, where it is for the "sole purpose of enabling the apprentice to perform the work for which they have been employed". Given that all apprentices are employed for the purposes of undertaking 'a job with training', which is the definition of an apprenticeship (DfE 2023, 28), the separation between 'work' and 'learning/training' seems unhelpful. This may be as a consequence of the funding rules being driven by and focused on establishing what is eligible to fund through the Apprenticeship Levy rather than quantifying amounts of required learning. However, it presents a particular challenge when seeking to establish the total learning/training hours required to complete an apprenticeship, which is required to establish a credit value.

The Challenge of Determining On- and Off-the-Job Learning/Training Hours

As already described, all learning/training hours required to complete the apprenticeship must be undertaken within 'normal working hours' (defined as the hours for which the apprentice would normally be paid but excluding overtime and statutory leave entitlement) with apprentices working sufficient hours each week to undertake regular training and any required on-the-job activity to ensure successful completion of the apprenticeship. This provides an upper ceiling for the total amount of learning hours required to complete an apprenticeship, as this must be no more than the total normal working hours. However, it does not indicate what amount of normal working hours should be dedicated to learning/training on the job or off the job.

The statutory requirement for a minimum amount of off-the-job learning/training hours that are to be undertaken within normal working hours is specified in the funding rules, and a method for calculating the minimum off-the-job training hours is provided. However, this does not equate to the amount of learning/training hours that may be required to develop the KSBs to achieve occupational or professional competence. Firstly, the funding rules specify a 'minimum' statutory requirement or entitlement, and employers and providers may agree that more off-the-job learning/training hours are required. Secondly, the funding rules historically have not specified the amount of on-the-job training/learning hours that may be required to contribute to the successful development of KSBs.

This means that while there is an upper limit to the amount of learning/training hours, provided by the amount of normal working hours and a statutory minimum amount of off-the-job learning/training hours, this does not establish the total amount of apprenticeship learning hours required to determine credit value.

As a consequence of the way that on-the-job training has been defined negatively as *not* off-the-job training "that does not specifically link to the knowledge, skills and behaviours set out in the apprenticeship" (DfE 2023, 38), there is a need to clarify the role of on-the-job training that does contribute to the development of the KSBs. The risk of not doing so makes it more likely that policymakers and practitioners will mistakenly conflate the statutory minimum off-the-job training entitlement with the total number of learning/training hours required to develop the KSBs set out in an apprenticeship standard. This mistaken interpretation further risks a gross underestimation of the amount of learning/training hours required to develop full occupational competence, particularly for higher and degree apprenticeships.

There is also a risk that the specified learning hours for a mandatory qualification within an apprenticeship standard can be conflated with the total learning/training hours required for the apprenticeship. While, as we will see, the numbers of learning hours required for a mandatory qualification can be significantly more than the minimum statutory off-the-job hours, this should not always be taken to mean that they equate to the total apprenticeship learning/training hours.

The Challenge of Determining Apprenticeship Learning/Training Hours for Different Kinds of Apprenticeships

Although the IfATE intend to integrate end point assessment (EPA) for apprenticeships with a statutory body/regulator, with the assessment arrangements determined by the regulator that typically reflect the requirement for entry into that occupation/profession, most apprenticeships currently are 'non-integrated'. There are two types of non-integrated apprenticeships: those that include a mandatory qualification and those that do not. For example, the Level 3 community health and wellbeing worker apprenticeship standard, like all apprenticeships, requires Level 2 functional skills (English and maths) qualifications but does not specify any other mandatory qualifications. End point assessment stands alone as the impartial test of occupational competence, and there are, as such, no mandatory qualification-related learning hours. However, many apprenticeships and the significant majority of apprenticeships in the healthcare sector, for example, do include the requirement to complete a mandatory qualification.

For these apprenticeships, mandatory qualifications must be successfully completed for apprentices to pass through the 'apprenticeship gateway' at the end of the practical learning period and be eligible to undertake the EPA that establishes full occupational or professional competence. As an example, the current approved version of the Level 3 senior healthcare support worker apprenticeship standard states: "You must complete a Level 3 diploma in healthcare support prior to taking the end-point assessment" (IfATE 2023). Similarly, degree apprenticeships include the completion of a specified degree as a mandatory qualification. For example, the Level 7 physician associate degree apprenticeship requires that all apprentices complete a diploma or MSc in physician associate studies (IfATE 2022). In both instances, the learning hours required to complete the mandatory qualification must be undertaken within normal employed hours and not in an apprentice's own time.

Typically, the mandatory qualifications specified within apprenticeship standards have a credit value and associated notional learning hours. For example, the diploma in healthcare support, required for the senior healthcare support worker apprenticeship has a value of 83 credits at Level 3 and 830 learning hours, while the MSc in physician associate studies has a value of 180 credits at Level 7 and 1800 learning hours. For non-integrated apprenticeships such as these, the mandatory qualification learning hours do not include any hours associated with undertaking EPA at the final stage of the apprenticeship.

For some statutory and regulated professions such as nursing, higher education and degree qualifications that have been accredited by the relevant statutory and regulated body, for example the Nursing and Midwifery Council (NMC) in the UK, now constitute integrated higher and degree apprenticeships. For this kind of apprenticeship, completion of the accredited mandatory qualification itself (along with required functional skills) is deemed to meet the requirement for end point assessment of occupational or professional competence. This means that the learning hours required to complete the accredited mandatory qualification also normally constitute the learning hours required to complete the apprenticeship. For example, the nursing associate foundation degree has a value of 240 credits at Levels 4 and 5 and requires 2400 learning hours, which also constitutes the total learning hours for the Nursing Associate apprenticeship.

However, the registered nurse integrated degree apprenticeship, like all other preregistration nursing degree programmes in the UK, currently requires a minimum of 4600 hours to be accredited by the NMC, which has a statutory responsibility to oversee the standards of nursing education. The NMC also require that a minimum of 2300 of these hours must be 'practice hours', which equates to on-the-job hours, undertaken in a healthcare workplace setting. This means that while the mandatory qualification within a standard integrated degree apprenticeship would require 3600 learning hours, the registered nurse degree apprenticeship requires an additional 1000 hours. Given that it is also required that 2300 of these hours are 'practice' hours, it can be assumed that at least 1000 of 'practice' (on-the-job) hours are in addition to the mandatory degree qualification learning hours.

Integrated degree apprenticeships that are not statutory or regulated professions have been designed with EPA embedded within degree programme modules delivered in the final stage of the apprenticeship and are required for the award of the degree and the apprenticeship certificate. For example, the healthcare science practitioner integrated degree apprenticeship has a value of 360 credits at Levels 4, 5 and 6 and 3600 learning hours.

The following examples illustrate the various relationships between normal working hours, the statutory minimum off-the-job learning/training entitlement, mandatory qualification learning hours, on-the-job learning/training hours and apprenticeship learning hours for different kinds of apprenticeships.

For non-integrated apprenticeships that do not have mandatory qualifications, consideration needs to be given to the amount of on-the-job learning/training hours that are required to develop the KSBs specified in the apprenticeship standard over and above the statutory minimum entitlement, as illustrated in Table 4.1. Once this is established, this can be added to the minimum off-the-job learning/training entitlement to

Table 4.1 Non-integrated apprenticeships without mandatory qualifications

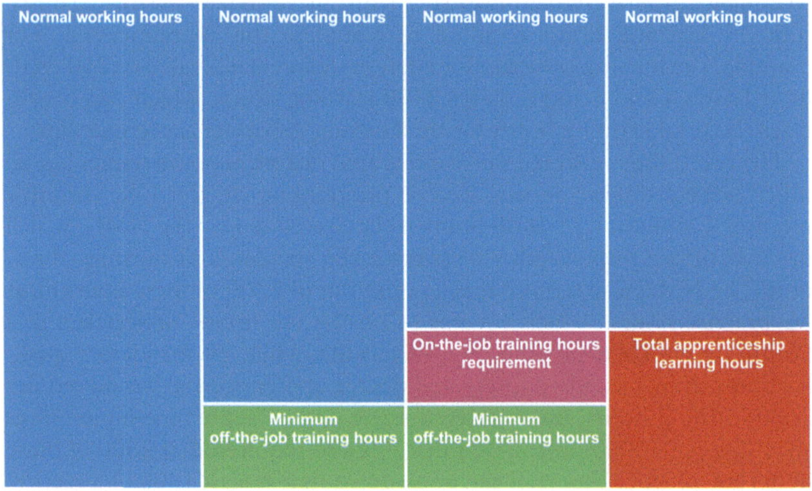

ascertain the total apprenticeship learning hours. A key risk for these kinds of apprenticeships is that the assumption is made that all the learning/training time needed to development the specified KSBs can be undertaken within the identified statutory minimum off-the-job hours. The risk arises from the conflation of the number of statutory minimum hours and the total hours required to develop the specified KSBs to achieve occupational or professional competence. 'Statutory minimum learning/training hours' and 'total required learning/training hours' are not the same thing.

For non-integrated apprenticeships with mandatory qualifications, consideration will need to be given to the relationship between the minimum off-the-job learning/training hours and the mandatory qualification learning hours, as shown in Table 4.2. In some cases, the mandatory qualification learning hours exceed the minimum off-the-job learning/training hours, which means that some of the mandatory qualification learning hours must be able to be undertaken on the job. In addition to this, there may be further on-the-job learning/training hours that are required to develop the KSBs specified in the apprenticeship standard. The combination of mandatory qualification learning hours and required on-the-job learning/training hours will indicate the total apprenticeship learning hours. This provides further evidence that 'statutory minimum learning/training hours' and

Table 4.2 Non-integrated apprenticeships with mandatory qualifications

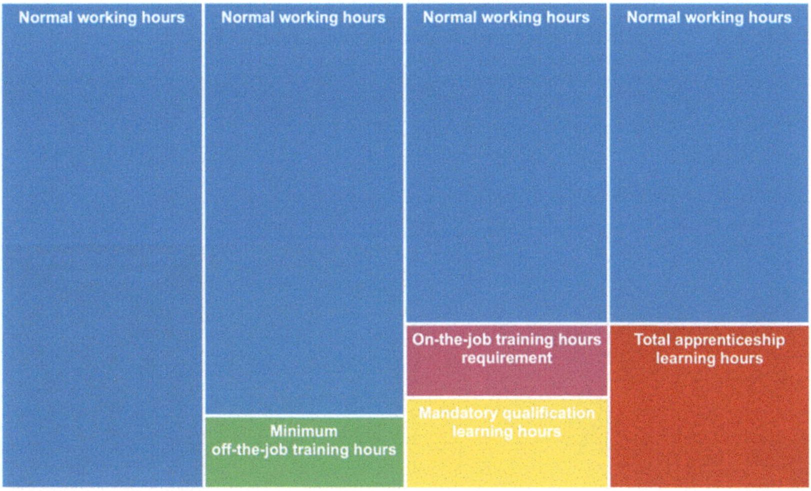

'total required learning/training hours' cannot be conflated as the latter is significantly larger than the former. It is vital that this central point is fully understood by policymakers to avoid the misguided conception that the learning/training hours for integrated higher and degree apprenticeships can be delivered within the statutory minimum learning/training hours.

For integrated higher and degree apprenticeships, the mandatory qualification will be a higher education qualification or degree that will have a high level of learning hours in relation to normal working hours, as demonstrated in Table 4.3. For example, 3600 learning hours for an integrated degree apprenticeship. This learning hours requirement will significantly exceed the minimum off-the-job learning/training entitlement and will necessitate that apprentices undertake a significant number of hours engaged in on-the-job learning/training. This means that the mandatory qualification learning hours must incorporate both on- and off-the-job learning/training hours to enable apprentices to develop the KSBs specified in the apprenticeship standard. This also means that the mandatory qualification learning hours normally equate to the total apprenticeship learning hours.

However, for some integrated degree apprenticeships, professional registration may require more learning/training hours than are specified by the degree qualification, such as registered nurse. In such cases, additional on-the-job training hours will need to be specified and the combination of

Table 4.3 Integrated higher and degree apprenticeships: Type 1

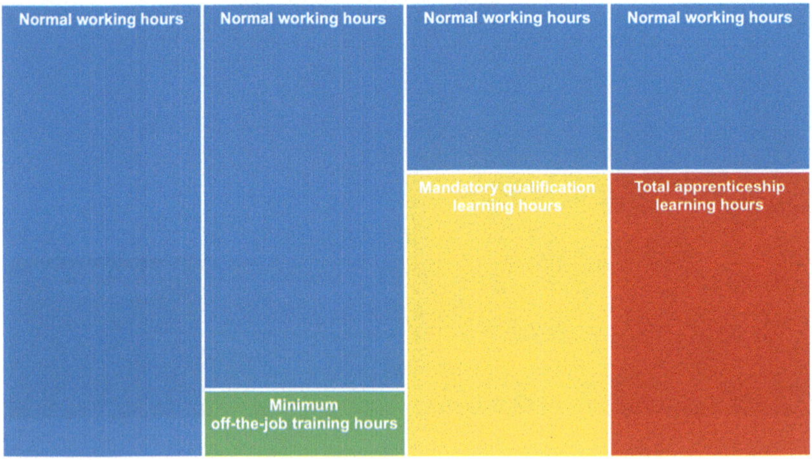

Table 4.4 Integrated higher and degree apprenticeships: Type 2

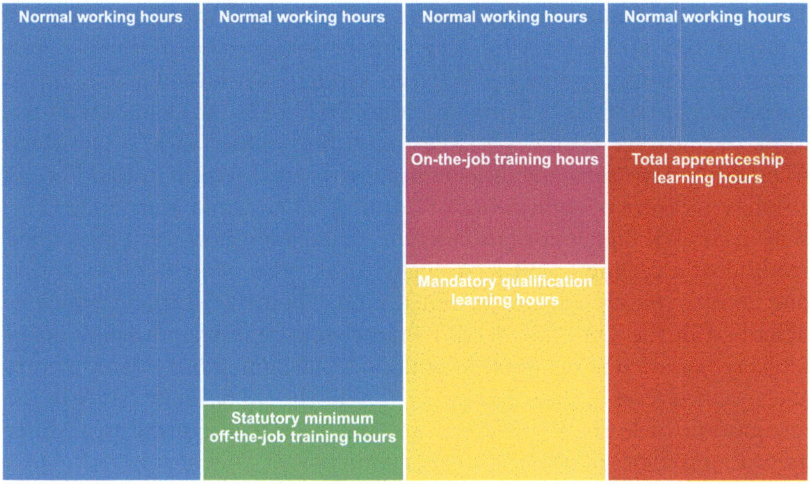

the mandatory qualification learning hours, and the additional on-the-job learning/training hours will comprise total apprenticeship learning hours.

For example, as indicated in Table 4.4, the registered nurse integrated degree apprenticeship does not have a separate EPA. Rather, successful completion of nursing degrees that are accredited by the NMC is deemed to be the means by which the requirement for demonstrating professional competence is met. In addition, to successfully complete the apprenticeship, apprentices will also have had to demonstrate the achievement of functional skills. While the mandatory registered nurse degree qualification will require 3600 learning hours, the NMC specifies that nursing degrees must include a minimum of 2300 'theory' hours and 2300 'practice' hours, totalling 4600 hours. This means that there is a requirement for an additional 1000 hours over and above the learning hours required to complete the mandatory integrated degree apprenticeship qualification. This indicates two things. Firstly, that the specified mandatory qualification learning hours do not always encompass all of the required learning/training time to successfully complete an integrated degree apprenticeship. Secondly, that the statutory minimum learning/training hours bear no resemblance to the amount of learning/training hours that are actually required to successfully complete an integrated degree apprenticeship.

Ascribing a Credit Value to Apprenticeships

The use of academic credit provides a common currency for learning and a mechanism for quantifying achievement irrespective of where it was completed "bridging the gap between formal, non-formal and informal learning opportunities" (Dunn and Marr 2022, 12). Academic credit is a form of "market currency" (Marr and Bravenboer 2017, 1), which whether used for entry to higher education or employment "should be, an absolute measure of equivalence of learning" (Dunn and Marr 2022, 22). Within the UK higher education sector, there is not however a prescribed method of ascribing credit to a qualification other than the notion that one credit typically equates to ten notional hours of learning which as described in the QAA HE Credit Framework for England definition of terms is "awarded in recognition of the achievement of learning outcomes at a specified level" (QAA 2021, 11). Credit should facilitate parity in the recognition of learning achievements, but how it is applied "will differ according to types of delivery, subject content and student cohorts" (QAA 2021, 11). A fundamental benefit of credit is that it can support progression for learners and pathways through educational levels, but its absence can create barriers for people seeking to progress in their careers.

When considering student admission to a course of study, higher education providers want to be assured that learners are adequately prepared for higher-level learning, which is viewed through the value of the prior qualification in terms of the level and volume of learning undertaken. Establishing the value of an apprenticeship in preparing learners for higher-level learning is relatively straightforward. Mapping the KSBs of an apprenticeship standard to the SEEC-level descriptors provides a clear articulation of the learning value of apprenticeships as a work-based programme, to support appropriate progression onto HE. As identified in the QAA HE Credit Framework for England (2021), SEEC-level descriptors are designed to enable learning that is described in the same way to gain equivalent recognition, so they are an HE sector-recognised tool to benchmark learning in apprenticeships. Using the SEEC-level descriptors is important to achieving consistency in approach to articulating the learning expectation of apprenticeships so that higher education providers understand their value.

Ascribing a credit value (or number of credits) to an apprenticeship is far more complex and made more challenging because there is not a prescribed method of ascribing credit to a qualification, other than the notion, as stated above, that one credit typically equates to ten notional hours of learning. Guidance presented within HE does not relate directly to apprenticeships and is instead based on the ascription of credit to qualifications and modules/units of study. This typically focuses on how credit is ascribed to units of study where the total credit value of the qualification is known. The whole qualification is then divided up into smaller elements (modules/units) in equal and/or various sizes, which are combined to comprise the whole qualification or learning programme. The challenge with apprenticeships however is that the overall size of the qualification is unknown, making the ascription of credit more complex. Whilst guidance within the HE sector does not provide a prescriptive method to achieve this, it does present key principles which can be used to develop a process of ascribing credit to apprenticeship standards.

The HE Credit Framework for England makes it clear that "credit assigned to modules or courses is based on the approximate number of hours a learner is expected to spend learning to achieve the learning outcomes for that module, including guided and independent study" (QAA 2021). When we relate this to apprenticeships, this would be the time required to develop the KSBs of the apprenticeship standard in order to achieve full occupational or professional competence. In the development of access to HE diplomas, it is made clear that the ascription of a credit value to each unit within the 60-credit qualification "is a matter of professional judgement for Access Validating Agencies (AVAs), exercised within their validation process (QAA 2020, 12). Again, these judgements are based on the concept of a 'notional learning hour', where one credit represents those learning achievements that can be demonstrated in 10 notional learning hours" (QAA 2020, 12).

Guidance provided by the Qualification and Curriculum Authority (QCA) consolidates this view. The amount of credit awarded for achievement of learning outcomes for a specified unit of study should be based on the notion that: learning time is defined as the time taken by learners at the level of the unit, to on average complete the learning outcomes to the standard determined by the assessment criteria; that one credit is awarded for learning outcomes achievable in ten hours of learning; that the credit value of each unit remains the same regardless of the method of assessment used or the qualification for which it contributes to (QCA 2008). Whilst the value of credit in relation to the number of learning hours is consistent with the HE Credit Framework for England, QCA make it clear that "the process to determine credit value cannot be prescribed" (2008,

30). The ascription of a credit value to a unit of learning is considered here to be part of "the holistic process of unit development", the guiding principles of which will be "similar to those relating to determining level" (QCA 2008, 30).

Core to ascribing a credit value is the notion of learning time but the QCA guidance made it clear that this "is not an exact, scientific measure but a judgement made and agreed by those who are informed and experienced in the relevant area of achievement and who understand the contexts in which the learning can take place" (QCA 2008, 28). There is therefore "no one agreed set of principles or methodology for estimating learning time...and there are no simple formulae" (QCA 2008, 30). That said, when seeking to arrive at a credit value through an estimation of learning time, "the credit value of a unit is arrived at by estimating the learning time and dividing it by 10" (QCA 2008, 31), which further reflects the one credit per ten hours of notional learning metric.

An important aspect of the process of ascribing a credit value to a unit of learning, as identified by QCA, is "the involvement of a panel or group of relevant experts" to ensure that the "credit value is recognised and accepted by relevant stakeholders" (QCA 2008, 30). This will mean that credit values will have been arrived at "through discussion between professionals as part of the overall unit development and validation process, drawing on expertise in subject areas" (QCA 2008, 30). In a similar vein to the QCA, guidance from the Scottish Credit and Qualifications Framework (SCQF) indicates that the allocation of SCQF credit points to qualifications is established by individuals who have expertise within the subject and make a professional judgement on the number of hours a typical learner would take to achieve the learning outcomes at a given level (SCQF 2017). In alignment with both the HE Credit Framework for England (QAA 2021) and QCA (2008), the time estimated is referred to as learning hours and one SCQF credit equates to ten notional hours of learning (SCQF 2017). The SCQF guidance makes clear that consideration of time spent on all activities that lead to the achievement of learning outcomes must take place when ascribing credit to a learning programme or qualification.

In trying to establish a consistent method to ascribe a credit value to apprenticeships, reference to relevant accredited qualifications can provide valuable benchmarks. For integrated higher and degree apprenticeships, this is very straightforward as they necessarily include mandatory higher education qualifications. The learning hours for the qualifications listed in the QAA Frameworks for Higher Education Qualifications of UK Degree-Awarding Bodies and referred to in the QAA HE Credit Framework for England are established. For example, 3600 learning hours for a bachelor's degree with honours also normally constitutes the total learning hours for a Level 6 integrated degree apprenticeship. There are instances where the correlation between the learning hours for a degree qualification and those required for an integrated degree apprenticeship do not correlate. Using the example of the registered nurse degree apprenticeship as already discussed includes an additional 1000 practice hours over and above the standard 3600 learning hours required by the degree qualification. However, benchmarking for non-integrated apprenticeships that include a mandatory qualification is far more complex. For this particular type of apprenticeship, we can be confident that the learning hours associated with the mandatory qualification may only partially contribute to the total learning hours required. For those that do not include a mandatory qualification, direct benchmarking against other relevant qualifications may not be available.

Benchmarking for non-integrated apprenticeships can however still take place, although indirectly. This can involve consideration of precedents currently available in the skills sector that potentially align with the total amount of apprenticeship learning hours or credit value being proposed. For example, the Universities and Colleges Admissions Service (UCAS) in England states and many other organisations providing information about apprenticeships advise that Level 3 apprenticeships are "generally considered to be the same level as two A level passes" (UCAS 2023). If an equivalent credit value for two A level qualifications could be ascertained or calculated, then, in principle, this could provide a helpful benchmark for non-integrated Level 3 apprenticeships.

When we consider relevant sector guidance and seek to ascribe a credit value to apprenticeships, by way of a reminder, the process needs to establish how many learning hours are required for an apprentice to achieve the KSBs, or the learning outcomes, specified in the relevant apprenticeship standard. We must also consider the notion as stated by IfATE (2022) that all apprenticeship standards require a combination of on- *and* off-the-job training.

Drawing together the key principles presented within relevant sector guidance, the process for ascribing credit to modules/units within a qualification can be articulated within five key steps:

1. Development of learning outcomes
2. Development of assessment criteria
3. Estimation of learning hours through discussion by experts within the relevant field
4. Benchmarking against precedents within the sector
5. Agreement and validation of credit value

This process cannot, however, be applied directly to apprenticeships without consideration of their specific regulatory requirements, including normal working hours, off-the-job learning/training hours, mandatory qualification learning hours and on-the-job learning/training hours.

By drawing on the key regulatory requirements for apprenticeships and applying them to the general process for ascribing a credit value to qualifications, it is possible to develop an apprenticeship specific process. The objective of the process will be to determine the total number of learning/training hours it will take for a learner to achieve the KSBs specified in the apprenticeship standard that are required to become occupationally or professionally competent. It is important to note that for approved apprenticeship standards the 'learning outcomes' are established and expressed as the required KSBs and that the 'assessment criteria' are described within the published apprenticeship EPA plan. This means that with regards to the general qualification process outlined above, the development of learning outcomes and assessment criteria for apprenticeship standards has already been completed and are, as such, redundant. However, additional steps will be required to appropriately reflect the relationships (expressed in hours) between different elements of the regulatory requirements of apprenticeship standards.

The specific process for ascribing a credit value to apprenticeships, outlined in Table 4.5, can be articulated in more detail as follows:

Step 1: Calculate the total normal working hours in relation to the duration of the apprenticeship standard. This can be calculated by determining the normal weekly working hours and multiplying by the number of weeks over the duration of the apprenticeship, less statutory leave entitlement. For example, 37.5 normal working hours per week with 5.6 weeks statutory leave entitlement equates to 1740 normal working hours per year, multiplied by the duration (in years) of the apprenticeship. This provides the total normal working hours 'envelope' within which all apprenticeship training must take place.

Table 4.5 Illustration of steps and calculations for the credit value of an apprenticeship using the Senior Healthcare Support Worker (SHCSW), Police Constable (PCDA) and Registered Nurse (RNDA) apprenticeship examples

Steps	*Calculations*	*SHCSW*	*PCDA*	*RNDA*
1	Total normal working hours	3480	5220	6960
2	Total minimum statutory off-the-job training hours	556	834	1112
3	Total normal working hours (minus)—Total minimum statutory off-the-job training hours	2924	4386	5848
4	Total mandatory qualification learning hours (if applicable)	830	3600	3600
5	Total normal working hours (minus)—Mandatory qualification learning hours	2650	1620	3360
6	Total estimated or required additional on-the-job training hours	400	0	1000
7	**Non-integrated apprenticeships with no mandatory qualification** Total minimum statutory off-the-job training hours (plus) + Total estimated on-the-job training hours	–	–	–
	Non-integrated apprenticeships with a mandatory qualification Total mandatory qualification learning hours (plus) + Total estimated on-the-job training hours	1230	–	–
	Integrated higher and degree apprenticeships Total mandatory qualification learning hours (plus) + any PSRB required additional on-the-job training hours	–	3600	4600
8	Total apprenticeship learning hours (divide) / 10 = credit value (rounded to nearest 10 credits)	120	360	460

Step 2: Calculate the total minimum statutory off-the-job learning/training hours in relation to the duration of the apprenticeship standard. This is calculated in accordance with ESFA Apprenticeship Funding Rules guidance by calculating the number of weeks over the duration of the apprenticeship, less statutory leave entitlement multiplied by six hours per week. For example, 46.4 working weeks per year multiplied by 6 off-the-job training hours per week equates to 278 hours per year over the duration of the apprenticeship (in years).

This provides the statutory minimum off-the-job hours entitlement. It does not provide any information directly regarding how many learning/training hours are required for apprentices to develop to required knowledge, skills and behaviours to become occupationally/professionally competent.

Step 3: Subtract the minimum statutory off-the-job learning/training hours from the total employed hours to identify the total number of normal working hours within which on-the-job learning/training can, in principle, take place. Establishing this is particularly relevant for non-integrated apprenticeships without a mandatory qualification as it provides the parameters within which any required training beyond the statutory minimum entitlement must take place.

Step 4: Identify the total learning hours specified by the mandatory qualification as applicable for non-integrated apprenticeships and integrated higher and degree apprenticeships with a mandatory qualification.

Step 5: Where applicable, subtract the total mandatory qualification hours from the normal working hours to identify how many normal working hours are available for non-qualification or additional required on-the-job training/learning to take place.

Step 6: Estimate the total amount of hours that it would take to complete the on-the-job non-qualification or additional required learning/training activities to support the development of the knowledge, skills and behaviours of the apprenticeship standard over the duration of the apprenticeship. To be clear, the estimated on-the-job learning/training hours identified are in addition to and not contained within the total learning hours specified by a mandatory qualification. This is important as some mandatory qualifications are designed to be 'work-based'.

If required additional on-the-job learning/training hours are not specified by a PSRB, this will require the identification of the range of learning/training activities that will take place to support the development of the KSBs of the apprenticeship standard. This might also include identification of KSBs that can only be developed on the job. For example, 'developing knowledge of local ways of working'.

This is a key step, particularly for non-integrated apprenticeships whether with or without a mandatory qualification. Categorising types of on-the-job learning/training will, to a degree, be sector specific, and it is advised that the process includes the formation of an expert employer panel to both sense check proposed on-the-job learning/training categories and provide information regarding typical patterns of engagement with agreed on-the-job learning/training activities. Specific feedback from an expert employer panel regarding the number of hours (per week or perhaps per year) that apprentices will typically engage in on-the-job learning/training activities will inform the estimation of the total on-the-job learning/training hours required.

Tables 4.6 and 4.7 provide an illustration of the approach used to gather information from the expert employer panel for the senior healthcare support worker apprenticeship where members of the panel were drawn from the healthcare apprenticeships Trailblazer Group and an example of an on-the-job training activity template (in this case for the senior healthcare support worker apprenticeship).

Where a PSRB has specified learning/training hours that are additional to the total learning hours required for a mandatory qualification, these additional hours can be identified. For example, where the NMC specifies that trainee nurses need to have completed 4600 hours in total to meet the requirements for registration, comprised of 2300 'theory' hours and 2300 'practice' hours, this identifies that there is a requirement for 1000

Table 4.6 Definition of terms used to determine total apprenticeship learning hours

Normal working hours	The hours for which the apprentice would normally be paid but excluding overtime and statutory leave entitlement. All apprenticeship learning/training must take place within 'normal' working hours.
Minimum off-the-job training hours	"Off-the-job training is a statutory requirement for an English apprenticeship...The minimum requirement, for apprentices working 30 hours or more per week is an average of 6 hours of off-the-job training per week (i.e., 20% of 30 hours) over the planned duration" (DfE 2023, 38–39).
On-the-job training hours	Training hours that are required to develop the KSBs of an apprenticeship standard that are over and above the minimum off-the-job training hours and the mandatory qualification learning hours.
Mandatory qualification learning hours	The total learning hours indicated for a specified mandatory qualification for an identified apprenticeship standard. Typically, also indicated by the credit value (or number of credits) for a qualification based on 10 learning hours per credit.
Total apprenticeship learning hours	The addition of either (a) the minimum off-the-job training hours and the identified on-the-job training hours required to develop the required KSBs for an identified apprenticeship standard or (b) the mandatory qualification learning hours and the additional on-the-job training hours required (where applicable) to develop the required KSBs for an identified apprenticeship standard

Table 4.7 On-the-job training activity template used for the Level 3 senior healthcare support *worker apprenticeship*

On-the-job training activity	Estimated hours per week	Estimated total days	Additional information
Orientation/Induction: Activities that introduce you to your role, working practices and working environment, such as organisation specific codes of conduct and procedures.			
Shadowing: Observing the working practice of an established member of staff for a period of time, to learn about relevant aspects of your job role.			
Co-working: Working together with another employee to carry out tasks and duties in the work environment that are relevant to your job role.			
Practice simulation: Practicing a real-life scenario or procedure through simulation before you undertake it with an actual patient/client.			
Demonstration: Watching a specific task or a particular process being carried out by an established member of staff to learn how to carry out the task/process yourself.			
Self-instruction: Taking time to find out about something or to practice a procedure yourself to help you develop the knowledge, skills and behaviours relevant to your job role.			
Delegation: Being assigned a task that would normally be carried out by a colleague to help you develop knowledge, skills and behaviours relevant to your job role.			
Job rotation: Where you engage in a range of job roles/activities to help you develop knowledge, skills and behaviours relevant to your apprenticeship.			
Other activities, please specify:			
Total			

hours in addition to the 3600 hours normally specified in the registered nurse degree apprenticeship qualification.

Step 7: Calculate the total learning hours for the apprenticeship.

For non-integrated apprenticeships with no mandatory qualification, the calculation requires adding the statutory minimum off-the-job training entitlement hours to the total estimated on-the-job learning/training hours. For non-integrated apprenticeships with a mandatory qualification the calculation requires adding the total mandatory qualification hours to the estimated (additional) on-the-job learning/training hours. For all non-integrated apprenticeships, it is advised that the proposed total number of apprenticeship learning hours proposed is 'sense checked' with the expert employer panel prior to confirmation.

For integrated higher and degree apprenticeships, the total apprenticeship learning hours will be the same as the total learning hours for the mandatory qualification, as is the case for the police constable degree apprenticeship. As identified in step 6 above, a PSRB may specify learning/training hours that are additional to the total learning hours required for a mandatory qualification. The proposed total apprenticeship learning hours can, at this point, be benchmarked against relevant precedents, including existing relevant qualifications.

Step 8: Once the proposed total apprenticeship learning hours are confirmed, divide by 10 to establish the credit value for the apprenticeship standard. This is based on the ten learning hours per credit metric specified within, for example the QAA Higher Education Credit Framework for England (QAA 2021).

Policy and Practical Impacts

Beyond the ascription of a credit value to the Level 3 senior healthcare support worker apprenticeship, the main focus is specifically on conveying the significant value the apprenticeship brings in terms of preparing learners for progression onto relevant HE programmes that lead to professional registration in nursing and allied healthcare roles. This is imperative because, at present, in England there is an absence of parity with learners who have other recognised Level 3 qualifications.

By ascribing a credit value, the value that is gained from undertaking the senior healthcare support worker apprenticeship can be articulated in a language that HE provider admission tutors and programme leaders can recognise. It also helps give a better understanding of, and recognition for, the unique and specific value of this particular apprenticeship and through that awareness of how more people could be taking this apprenticeship to progress to HE programmes leading to professional registration in nursing, midwifery, dietetics, diagnostic radiography, occupational therapy, physiotherapy, podiatry and speech and language therapy as examples. Given that the UK's university admissions body is presently seeking to ascribe UCAS tariff points to apprenticeships in the same way that post-16 Level 3 qualifications—such as A levels, T levels, BTECs and Cambridge Technical (Applied Generals)—are ascribed points to help determine entry criteria for university courses, there is even greater urgency to achieving some national consistency (Noble 2023).

There are however lots of pitfalls and potential inconsistency in any process which ascribes UCAS tariff points as university entry requirements differ greatly across the UK. Such variability promotes inconsistency nationally and fails to benefit those who are seeking to progress their careers. In terms of the senior healthcare support worker apprenticeship, there are also challenges in the way in which the mandatory qualification is recognised. Given it is a pass/fail qualification, the UCAS tariff points allocated would be low in comparison to other Level 3 awards. It does not assist if the absence of UCAS tariff points means that the system defaults to the qualification within it, because it diminishes its value and implies it is not an effective preparation for learners progressing to HE. Equally, it fails to assist if there is then a focus on the mandatory qualification, as the number of hours of learning will still be low compared to a BTEC, for example. This will lead some admission tutors, when making that comparison, to conclude that the apprenticeship is of 'less value'. It is these barriers that this chapter aims to address in order to make progression routes clear and explicit and at the very least ensure the UK has the healthcare service it needs and deserves.

There is no doubt that something transformational is needed in the APEL policy space, and at a national level of coordination, to shift the narrative around credit not just for all-important public sector job roles but to support career ambitions in other professions and industrial sectors. Currently, individuals are reliant on an HE system that differs from institution to institution, provider to provider. What is needed is a joint further and higher education credit framework for England that is binding. The role of any such framework is important because it can structure recognition of credit on a national basis and use credit to empower a process of enabling people to gain recognition and progress. To date, the reliance on university autonomy and a higher education provider's own policy of recognition falls short of every learner having their prior learning recognised and remains fundamentally unfair. Empowering the way in which credit is and can be used will, inevitably, provide increased flexibilities and greater access to provision that supports lifelong learning.

Positioning On-the-Job Learning

Research in apprenticeships in England have referenced and placed emphasis on the transformatory aspects of teaching and learning through *on-the-job* training such as "to think and work outside the box" (Rowe et al. 2017, 187) and through increased individual activities such as "action learning, critical reflection, investigation" (Lester et al. 2016, 23). ESFA/DfE apprenticeship funding rules have, more recently, removed all references to on-the-job training other than in the context of describing what off-the-job training *is not* and in being more explicit, by way of a funding rules definition, that it is learning done 'outside' of the apprenticeship (DfE 2023 125). Such a move is at odds with an earlier impetus for the government when it published its 'Plans for Growth', which, when announcing 10,000 new advanced and higher apprenticeship places, stated:

> higher apprenticeships provide a good alternative to full-time education for young people who want to develop a career through on-the-job training. (HM Treasury 2011, 85)

Crucially this highlights a fundamental challenge in recognising the number of hours required in an English apprenticeship to deliver full occupational competence given the absence of any statutory or regulatory underpinning for conceptualising, developing or defining on-the-job learning/training alongside that delivered as fundable and, therefore, valued as 'off-the-job' training. In apprenticeship terms, off-the-job learning activities are reasonably well specified and validated for funding and regulatory purposes and can include the teaching of theory, practical training, learning support, time spent on assignments and revision, for example. On-the-job training is less well defined. *Indeed*, the employment website promoted as the number one job listing site worldwide sets out in its Career Guide (2022) the on-the-job training advantages which range from professional supervision, target-specific learning and a safe space to learn to time-saving benefits as individuals acquire specialist knowledge. The on-the-job training examples it gives, including, co-worker training, shadowing, delegation, job rotation, real-life scenarios and self-instruction, align well with Billet's suggestion (2016) that learning on-the-job is mostly predicated "on mimesis: observation, imitation and practice", which is all important to how individuals acquire socially derived knowledge required for occupational competence.

Conclusion

Unlike the England system, for the majority of European apprenticeship systems of delivery, apprenticeship learning/training *is* conceptualised in terms of both on- and off the job (Brockmann et al. 2020, iv) with formally agreed and negotiated on-the-job activities. Without giving equal priority to both forms, the current apprenticeships funding and delivery model in England—with an off-the-job requirement based on a minimum 20% of an apprentice's normal working hours over the planned duration—will remain rooted in a historic mode of delivery based on one-day-a-week or block release model rather than a delivery model that reflects employer, individual and economic needs of the 2020s. To reflect the required practice needed in the 2020s would necessarily involve how on- and-off-the-job learning is integrated, how off-the-job learning supports practice in the workplace and the development of competence, and how self-directed learning can be used in apprenticeships. The failure to sufficiently consider how on-the-job training is supported and integrated with off-the-job learning is a substantial omission. This, after all, is the central feature of any apprenticeship programme in England. Previous and current apprenticeship funding models reflect an out-of-date vision of apprenticeships in which providers deliver off-the-job learning that is then exemplified and applied in the workplace. In reality, apprenticeships are about employers and providers working in partnership, and providers need to undertake considerable activity to integrate on- and off-the-job learning and support on-the-job learning. As Lester, et al. propose (2016, 10), the workplace is more than just a place for gaining experience and a site of application for what has already been learned; it is, absolutely, an environment where knowledge can also be acquired. Therefore, providers must not focus just on the statutory minimum. The three-way engagement between employer, provider and apprentice is key to understanding that all parties have a responsibility to ensure that the whole apprenticeship (on- and-off-the job) constitutes a high-quality preparation for meeting apprenticeship standards. Contextualised within wider skills policy in England, there are further implications for the development of a more consistent FE and HE credit system of recognition and translation. It has the potential to support further research into the recognition of prior learning as a specialised pedagogy and for reflecting on practice in other professions.

At present, though, there remains little understanding or recognition amongst apprenticeship policymakers and regulators about methods that can be used for on-the-job learning in terms of what and how individuals learn in work. This is despite expectation within all professions, and especially in nursing, that learning is continuous, and, with changes in working practices and workplace technologies, a growth in the complexity of the knowledge, skills and behaviours is necessarily implied (Berings et al. 2007, 442). Finally, English apprenticeships are an education and productivity success story. With the stretch from those traditional craft and trade occupations to technical, managerial, associate professional and professional job roles, the development of new work-based, work-integrated progression pathways includes doing what they were intended to do: increase productivity and support social mobility by opening up access routes to the professions, where apprenticeships are seen as an aspirational choice for young people and adults of all ages, at all stages in their career and from all backgrounds (Crawford-Lee and Wall 2018, 238).

In the next chapter, the issue of access to opportunities in apprenticeships and progression from apprenticeship into the next level of learning continues to be explored as the growth of higher and degree apprenticeships have led to accusations of a 'middle-class landgrab' with some degree apprenticeship opportunities compared to Oxbridge admission in terms of competitiveness. Apprenticeships are, of course of significant value for many young people. Their use should be supported and championed, but so too should apprenticeships for older workers. In its 2020 vision for apprenticeships (DBIS 2015a), Westminster government was clear that apprenticeships should be an attractive offer for young people and adults leading to an apprenticeship system, in England, that champions an all-age programme.

References

Anderson, A., and M. Crawford-Lee. 2016. The Role of Higher Education in a World-class Apprenticeship System. In *A Race to the Top: Achieving Three Million More Apprenticeships by 2020*, ed. David Way. Winchester: The Winchester University Press.

Anderson, A., D. Bravenboer, and D. Hemsworth. 2012. The Role of Universities in Higher Apprenticeship Development. *Higher Education Skills and Work-based Learning* 2 (3): 240–255.

Berings, M., R. Poell, and J. Gelissen. 2007. On-the-job Learning in the Nursing Profession: Developing and Validating a Classification of Learning Activities and Learning Themes. *Personnel Review* 37 (4): 442–459.

Billett, S. 2016. Promoting Learning Through Work-based Experience: Mimetic Learning in Action. In *Facilitating Work-Based Learning: A Handbook for Tutors*, ed. R. Helyer, 247–263. London: Palgrave Macmillan.

Bravenboer, D. 2016. Why Co-design and Delivery is a "no-brainer" for Higher and Degree Apprenticeships Policy. *Higher Education Skills and Work-Based Learning* 6 (4): 384–400.

Bravenboer, D., and S. Lester. 2016. Towards an Integrated Approach to the Recognition of Professional Competence and Academic Learning. *Education + Training* 58 (4): 409–421.

Brockmann, M., Laurie, I., Smith, R., and Cui, V. 2020. *On- and Off-the-Job Training in Apprenticeships in England*. Available at: https://www.gatsby.org.uk/uploads/education/reports/pdf/onandoffthejobtrainingv5.pdf (Accessed 31 July 2023).

Crawford-Lee, M. 2022. *Gamekeeper Turned Poacher: Discursive Shifts in Apprenticeship Reform in England, 2011–2021*. Thesis (DProf). Middlesex University. London: Middlesex University

Crawford-Lee, M., and T. Wall. 2018. Sustainability 2030: A Policy Perspective from the University Vocational Awards Council. *Higher Education, Skills and Work-based Learning* 8 (3): 233–242.

Department of Business Innovation and Skills (DBIS). 2011. *Specification of Apprenticeship Standards for England*. London: BIS.

———. 2013. *Specification of Apprenticeship Standards for England*. London: BIS.

———. 2015. *The Future of Apprenticeships in England: Guidance for Trailblazers—From Standards to Start*. London: Stationery Office.

———. 2015a. *Apprenticeships (in England): Vision for 2020*. Available at: https://assets.publishing.service.gov.uk/media/5a74eacde5274a59fa715f9d/BIS-15-604-english-apprenticeships-our-2020-vision.pdf (Accessed 1 June 2024).

Department for Education (DfE). 2019. *Independent Panel Report to the Review of Post-18 Education and Funding, (the Augar Report)*. DfE, London.

———. 2021. *Skills for Jobs: Lifelong Learning for Opportunity and Growth*. DfE, London.

———. 2022a. *Skills and Post-16 Education Act*. London: The Stationery Office.

———. 2023. *Apprenticeship Funding Rules August 2023 to July 2024 (v1)*. Available at: https://assets.publishing.service.gov.uk/government/uploads/system/uploads/attachment_data/file/1155957/Apprenticeship_funding_rules_2324_Version_1.pdf (Accessed 1 July 2023).

Department for Education and Employment (DfEE). 1997. *National Committee of Inquiry into Higher Education (Dearing Committee)*. London: Stationery Office, (ED 266).

Dunn, C., and L. Marr. 2022. Credit Where Credit's Due: The Role of the Open University in Credit Recognition. In *Widening Access to Higher Education in the UK: Developments and Approaches Using Credit Accumulation and Transfer*, ed. W. Turnbull and H. Woolf, 11–22. Open University Press, London.

Health Education England (HEE). 2014. The Talent for Care: A National Strategic Framework to Develop the Healthcare Support Workforce. *HEE*. Available at: https://www.hee.nhs.uk/sites/default/files/documents/TfC%20National%20Strategic%20Framework_0.pdf (Accessed 4 August 2023).

HM Treasury, BIS. 2011. *Plan for Growth*. HM Treasury, London. Available at: https://assets.publishing.service.gov.uk/media/5a7cb81e40f0b6629523b637/2011budget_growth.pdf (Accessed 23 July 2023).

IfATE. 2023. *Senior Healthcare Support Worker Apprenticeship Standard, IfATE*. Available at: https://www.instituteforapprenticeships.org/apprenticeshipstandards/senior-healthcare-support-worker-v1-4 (Accessed 4 August 2023).

Indeed. 2022. *On-The-Job Training Examples (With Benefits and Tips)*. Available at: https://uk.indeed.com/career-advice/starting-new-job/on-the-job-training-examples (Accessed 2 June 2023).

Institute for Apprenticeships and Technical Education (IfATE). 2022. *Degree Apprenticeships, IfATE*. Available at: https://www.instituteforapprenticeships.org/developing-new-apprenticeships/degree-apprenticeships/ (Accessed 4 August 2023).

Jeffrey, C. 2016. *Is Your Vision the Same or Similar to Mine – Can You See What I See?, Association of Employers and Learning Providers*. Bristol. Availablec at: https://www.aelp.org.uk/news/aelp-blog/details/is-your-vision-the-same-or-similar-to-mine-can-you/ (Accessed 26 July 2017).

Leitch, S. 2006. *Prosperity for All in the Global Economy—World Class Skills*. London: Stationery Office.

Lester, S. 2016. Practice as Research: Developing the Workplace Project. In *Contemporary Approaches to Dissertation Development and Research Methods*, ed. V. Storey and K. Hesbol, 117–133. Hershey, PA: IGI-Global.

———. 2020. Creating Conditions for Sustainable Degree Apprenticeships in England. *Higher Education, Skills and Work-based Learning* 10 (5): 701–714.

Lester, S., D. Braveboer, and N. Webb. 2016. *Work-integrated Degrees: Context, Engagement, Practice and Quality*, A Literature Review for QAA. Gloucester: QAA.

Marr, L., and D. Bravenboer. 2017. Editorial: 'widening participation, lifelong learning and the role of higher education credit'. *Widening Participation and Lifelong Learning* 19 (3): 1–5. https://doi.org/10.5456/wpll.19.3

NAS (National Apprenticeship Service). 2011. *Higher Apprenticeship Fund Prospectus*. Coventry: NAS.

NHS. 2019. The NHS Long Term Plan. *NHS*. Available at: https://www.longtermplan.nhs.uk/wp-content/uploads/2019/08/nhs-long-term-plan-version-1.2.pdf (Accessed 4 August 2023).

Noble, J. 2023. UCAS Points for Apprenticeships Planned by the End of 2023. *FE Week*. Available at: https://feweek.co.uk/ucas-points-for-apprenticeships-planned-by-the-end-of-2023/ (Accessed 23 June 2023).

PARN (Professional Associations Research Network). 2015. *Professional Bodies and Apprenticeships: A Report for the Gatsby Charitable Foundation*. Bristol: PARN.

Quality Assurance Agency (QAA). 2020. The Access to Higher Education Diploma Specification. Available at: https://www.qaa.ac.uk/docs/qaa/guidance/access-diploma-specification-20.pdf?sfvrsn=aeb3cb81_4 (Accessed 1 September 2022).

———. 2021. *Higher Education Credit Framework for England: Advice on Academic Credit Arrangements*. Available at: https://www.qaa.ac.uk/the-quality-code/higher-education-credit-framework-for-england (Accessed 4 October 2022).

Quality Assurance Agency for Higher Education. (QAA) (2008). *Higher Education Credit Framework for England: Guidance on Academic Credit Arrangements in Higher Education in England*. Gloucester: QAA.

Qualifications Curriculum Authority. 2008. Guidelines for Writing Credit-Based Units of Assessment for the Qualifications and Credit Framework, (v3). Available at: https://dera.ioe.ac.uk/id/eprint/8626/2/Guidelines_for_writing_credit_based_units_4th_proof_web_ready.pdf (Accessed 4 October 2022).

Richard, D. 2012. *Review of Apprenticeships*. London: School for Startups.

Rowe, L., D. Moss, N. Moore, and D. Perrin. 2017. The Challenges of Managing Degree Apprentices in the Workplace: A Manager's Perspective. *Journal of Work-Applied Management* 9 (2): 185–199.

Scottish Credit and Qualifications Framework. 2017. SCQF Credit Points Explained: Notional Learning Hours. Available at: https://scqf.org.uk/media/bcehy5u3/credit-point-explained-march-2017-web.pdf (Accessed 4 October 2022).

SEEC. 2021. Credit Level Descriptors for Higher Education. *SEEC*. Available at: https://seec.org.uk/wp-content/uploads/2021/05/MDX_SEECDescriptors_Update-May-2021_Version-2_For-screen_AW13885.pdf (Accessed 2 August 2023).

UCAS. 2023. Level 3 Apprenticeships (Advanced Apprenticeships). *UCAS*. Available at: https://www.ucas.com/apprenticeships/england/level-3-advanced (Accessed 15 September 2023).

CHAPTER 5

Higher and Degree Apprenticeships: The Middle-Class Land Grab?

Mandy Crawford-Lee

An Initial Standpoint

Apprenticeships in England are an educational and productivity success story. With the stretch from traditional craft and trade occupations in advanced and intermediate apprenticeships to technical, managerial, associate professional and professional job roles, via higher and degree apprenticeships, the development of new work-based progression routes are doing what they were intended to do: increase productivity and support social mobility by opening up routes to the professions for under-represented and underserved cohorts where apprenticeships are seen as an aspirational choice for young people and adults from all backgrounds (Crawford-Lee and Wall 2018, 238). Yet there are those who make a moral claim about what the purpose of apprenticeships in England should be, today. Namely, that apprenticeships as defined by the Specification of Apprenticeship Standards for England (SASE) (Department for Business, Innovation and Skills [DBIS] 2009a) must remain synonymous with further education (FE) delivered by colleges and independent training

M. Crawford-Lee (✉)
University Vocational Awards Council, Bolton, UK
e-mail: mc15ext@bolton.ac.uk

J. Baldwin et al. (eds.), *The Future of Technical Education*,
https://doi.org/10.1007/978-3-031-70756-8_5

providers (ITPs). It is such highly principled, and at times amplified, views that are equally wedded to the idea that apprenticeship must remain a social inclusion route solely for the young and the unqualified (Crawford-Lee 2019).

Apprenticeships have, in the recent past, become a totem for 'purists' of the England skills system who support and perpetuate a variety of myths and opinions on the grounds that 'proper' apprenticeships should be designed to specifically meet the needs of underserved young people or adults furthest away from accessing first jobs and labour market opportunities. Indeed, higher and degree apprenticeships have been described as 'too expensive', diverting funding from young people who need lower-level skills programmes, 're-packaged graduate schemes' abused by employers who rebadge existing staff as apprentices (see Dawe 2019, 22; Gravatt 2019, 10; Augar 2019, 152), that they are not really in the spirit of apprenticeships at all, (Richmond 2020, 2) and in one report represent a 'middle-class land-grab' (House of Lords 2018, 75). A theme that has often repeated itself since (Cavaglia et al. 2020; Woolcock 2023).

Such striking opposition is intended to illustrate how differing descriptions of apprenticeships in England relate to their purpose and use and demonstrate how descriptions of higher and degree apprenticeships are operating within the wider higher education (HE) and skills discursive terrain. There is no denying that there are those who vocally oppose higher and degree apprenticeships and make claims as to their 'proper' quality or economic value. Often, the description of higher and degree apprenticeships is contrasted with a description of what apprenticeships have historically been or even what 'they should be'. This chapter is concerned with how the idea of higher and degree apprenticeships has been constructed within official HE and national skills policy discourse. It is not concerned with attempting to discover what kind of description of apprenticeships, specifically higher and degree apprenticeships, is true or false, or determine what the fundamental principles of apprenticeships are or should be. Rather, descriptions of the emergence, role and purpose of higher and degree apprenticeships in England are 'implicated' in the 'discursive struggles' (Bravenboer 2009, 10), or discursive shifts, leading to a determination of which descriptions are operating and have dominance in the field.

The Origins of the Policy

What follows is the historical context that frames the introduction of higher and degree apprenticeships and present some interpretations on the control of discourse in apprenticeship policy. By the time the Westminster coalition formed in 2010, there had been a decade or more of UK government encouragement to increase university attendance, primarily through a variety of widening participation policies and initiatives, (see Department for Education and Employment [DfEE] 1998; Department for Education and Skills [DfES] 2003) coupled with a 'global surge in the number of young people going to university' (Coughlan 2011). For some, such 'normalisation' of HE after half a century or more of expansion—in numbers of students, staff and institutions—at the beginning of the twenty-first century reflects wider changes to the economic and social context (Bathmaker 2003, 172), while for others it perpetuated the belief that a degree is more valuable than vocational training (Wall and Perrin 2015) given that technical and vocational training had long been regarded as inferior to academic and professional attainment and therefore that knowing is more highly regarded than doing (Gribble et al. 2015). This is no less true of arguments applied to the status of apprenticeships in England.

The variety of misconceptions and ideologies that create the centuries-long-held view of the academic and vocational divide, one that sees headlines such as 'degrees for the rich, apprenticeships for the poor' (McInerney 2018) and described in a study by Wall and Perrin (2015) (cited in Crawford-Lee and Wall 2018, 235), sit alongside an established societal belief, particular among the middle classes, that a degree has an intrinsic (and extrinsic?) higher value than an apprenticeship; that '[…] choosing a path other than university is a mark of failure' (Bawden 2011). The emergence of a higher-level skills debate by government that positions universities as central to UK economic recovery (Department for Business, Innovation and Skills [DBIS] 2009b, 2010; Cable 2010, 2011) and challenges them to demonstrate and justify their 'world-class' status occurs at the same time as the National Strategic Skills Audit (UK Commission for Employment and Skills [UKCES] 2010a, b); which highlights the need for apprenticeships to support further learning at HE level, while at the same time describing the purpose of higher education institutions (HEIs) as providing education and not 'training' (UKCES 2010a). This presents as an early challenge around balancing higher education's purpose with that of apprenticeships.

The nexus between universities and apprenticeships started to emerge under New Labour with the promise to pilot honours and master's programmes centred on the principles of the apprenticeship, namely employed status, technical expertise and occupational competency (DBIS 2009b, 6), as part of an 'expanded apprenticeship system'. Formal dialogue about the lack of progression from apprenticeship programmes to HE started to emerge in early 2010 (but had been discussed as early as 2005). Framed by a debate about equity and social mobility, the Lifelong Learning Network—a joint initiative in the UK between the Higher Education Funding Council for England (HEFCE), the Learning and Skills Council (LSC) and the Department for Education and Skills (DfES)—published a policy paper that raised the idea for the first time of a 'missing link' in the apprenticeship system in England. The paper was informed by three pieces of research written by the University Vocational Awards Council (UVAC): firstly, on progression from vocational and applied learning to HE across the UK on behalf of the UK Commission for Employment and Skills (UKCES) (2010c); secondly, apprenticeships (Anderson and Hemsworth 2005); and thirdly, vocational progression (Carter 2009) carried out on behalf of Westminster government. Each research contributed to the coalescing of the debate about the limits of the existing apprenticeship system in terms of breadth, prestige and opportunity and argued that the lack of progression to HE of apprentices was not due to the University and Colleges Admissions Service (UCAS) tariff points, articulation or barriers to admission; it was more to do with the lack of flexible part-time, work-based, higher education pathways—like higher apprenticeships—for them to progress on to:

> there is a lack of flexible, part-time, discrete higher education pathways open to these learners who made a decision [...] to take up a work-based learning route rather than a full-time academic route. To put such higher apprenticeship routes in place has the potential to dramatically open up access to the professions to young people from lower socio-economic groups, who are more than twice as likely to undertake vocational qualifications as those with parents in professional occupations. Higher apprenticeships are the missing link in the apprenticeship family, the component that provides for apprenticeships, parity of esteem with academic routes. Without clear work-based progression pathways, leading in appropriate sectors to professional or

para-professional accreditation, apprenticeships and advanced apprenticeships will always be seen as having a ceiling at level 3, only leading to technician occupations. (Hall et al. 2010, 2)

By 2010, apprenticeships at the higher level were not new; but they were in a very small number of occupational areas and industries. Apprenticeship frameworks comprising National Vocational Qualifications (NVQs) or NVQ units up to Level 4, functional skills, and an HE 'technical certificate', typically a foundation degree, were offered in engineering, ICT, telecommunications and accountancy (at the technician level). Higher apprenticeships with 850 starts in 2009 were considered 'a significant programme' by a non-departmental government body (NDGB), the National Apprenticeship Service (NAS), at a time when apprenticeship starts at all levels were growing rapidly from 280,000 in the academic year 2009–2010 to 457,000 in 2010–2011 (DBIS 2016), dominated by opportunities in low-level job roles in, for example, health and adult social care, business administration and customer service. NAS's commitment to consider the development of new higher apprenticeships at Qualification and Credit Framework (QCF) Levels 5, 6 and 7 was first mooted in 2009, to be developed in consultation with professional bodies and sector skills councils (SSCs). This development was positioned as enabling individuals with the aspiration and ability to progress from technician jobs to professional and possibly chartered job roles with an employer, thereby providing a work-based progression route comparable to the established and traditional 'academic' undergraduate route. Much of the argument for the implementation of higher apprenticeships was about freeing up HE pathways that both apprentices and their employers understood: programmes that were designed for people in work that combined technical knowledge with work-based competence.

The Age of Apprenticeships

In February 2011, the age of higher apprenticeships was described as having arrived (Joslin 2011, 1) when the latest Specification of Apprenticeship Standards for England (SASE) (Employment and Training 2011) was published and included for the first time the criteria and specification for higher apprenticeships at Levels 4 and 5. Following this, in March 2011, to accompany the spring budget the government published 'Plans for Growth', which announced 10,000 new advanced and higher apprenticeship places and stated 'higher apprenticeships provide a good alternative to full-time higher education for young people who want to develop a career through on-the-job training' (HM Treasury 2011, 85).

In what can be described as the start of a period of constant policy in 'motion' at the national level of government, concerning education and skills, in the same way the Leitch Review of Skills (Leitch 2006) was testimony to the way in which the UK's overarching policy goals were framed in relation to catching up and beating competitor Organisation for Economic Co-operation and Development (OECD) nations, 'New Challenges, New Chances: Further Education and Skills System Reform Plan' quickly followed on from SASE and Plans for Growth. Combined, they established the systems architecture for apprenticeships in England that we have today. *New Challenges, New Chances* laid out a government ambition to provide a ladder of opportunity through apprenticeships to 'clear and flexible progression routes to Higher Vocational Education' (DBIS 2011, 3). In the same year, Westminster government demonstrated its intentions practically with the launch of the Higher Apprenticeship Fund (HAF). In providing £25M to boost the development of 10,000 higher apprenticeship places (NAS 2011), it looked to stimulate the creation of new apprenticeship programmes. Seemingly too in recognition that the lack of a higher apprenticeship route might not just be a problem in terms of ensuring employees learn the right way but that it might also be limiting opportunity and social mobility based on the findings that half of all apprentices completing an advanced apprenticeship (at Level 3) showed interest in progressing to a degree-level equivalent course (UKCES 2010c, 34), yet failed to progress.

From a policy perspective, there was a dual objective in using the HAF for the development of a 'new wave' of apprenticeships. For employers, it was to address the challenge of growing skills at the higher levels, and for apprentices to create access to new employment opportunities with greater career progression options and a route to professional recognition and status (Crawford-Lee 2012, 285). An additional intent to investing in the expansion of higher apprenticeships was to change the popular perception of apprenticeships in England. Industries traditionally associated with apprenticeships (in engineering, construction and manufacturing) were to be joined by sectors associated with the professions in the hope that as higher apprenticeships grew in numbers, in sector coverage, in recognition and in importance to business and to individuals, the artificial division in terms of 'parity' of recognition between traditional academic disciplines and higher-level learning provision designed to raise skills levels in the workplace could be challenged and reduced (Crawford-Lee 2012, 286).

It is this last point that became central to the discursive narrative of apprenticeships in HE, as a form of work-based learning, in England from 2011 onwards. Given the strong tendency to see work-based learning as identified with low-status vocational qualifications or not appropriate to higher-level learning at all, HE and vocational training have been long considered as alternative sides of the English education system, characterised as 'a bifurcated system in which either narrow, specifically occupational qualifications, or traditional academic qualifications are offered' (Wolf 2011, 74). By this construct, apprenticeships were seen as the vocational option and A levels, followed by an honours degree, as the conventional, linear, route to employment, post-graduate. The reality was far more nuanced. HE has always had vocational elements represented by its long association of training for the professions where the minimum of a degree is required to practice but, like apprenticeships, the popular perception of HE has also been misconstrued and reflects how:

> The change in the size and character of the publics who have an interest in higher education and exert an influence on higher education policy greatly influences the nature and content of the discussions about higher education; particularly, who takes part in them, and the decisions that flow out of them. (Trow 2007, 25)

It is ironic that at the same time as more UK universities were welcomed into membership of the Russell Group—the most highly selective, research-intensive association that counts Oxford and Cambridge among its membership (Shepherd 2012)—HE was becoming more vocational than ever. Having come under pressure to improve the employability of students, this led to the substantial development of higher-level vocational programmes in universities and an expansion in the range of vocational programmes beyond those recognisable 'first generation' professional subjects in architecture, engineering, law and medicine to 'second generation' subjects such as business administration and social work (Guile and Evans 2010, 5). Working with employers honed HE's response to the development and/or delivery of work-integrated learning opportunities relevant for the modern-day jobs market, resulting in "programmes of a bespoke negotiated, adaptable and often multi-disciplinary nature" (Helyer 2015, 20). The association with and move to align non-technical apprenticeships with a funding and delivery mechanism for bachelor's degrees and master's programmes was still a number of years away but the revision to SASE in 2011, which put apprenticeship standards on a statutory basis for the first time and provided for apprenticeship programmes up to and including level 5, coupled with the HAF (NAS 2011), became watershed moments for the role of universities in higher apprenticeship development: as much for what they failed to address as they succeeded to initiate. Bravenboer (2016) describes the bifurcation of higher-level skills policy, which contributed to a low level of engagement by HE institutions in higher apprenticeships and the higher-level skills policy area, more generally (Anderson et al. 2012, 241). Despite government funding, it was the lack of alignment between the second SASE (DBIS 2011) and HE qualifications which operated as the major brake on institutional engagement.

Anderson et al. (2012) and Bravenboer (2016) are clear about the inhibitors. These included the fact that higher apprenticeships could not extend above Level 5 (or beyond sub-degree level); the size variability of the qualifications that could be delivered; the separation of academic 'knowledge' and technical 'competence' in the development of higher-level skills through separate qualifications or assessments; and the lack of an even-playing field given the absence of any public funding for prescribed HE qualifications that were included in delivery. Despite in principle support from professional bodies, the restriction on qualification level meant that higher apprenticeships could not be used to support learners' progression to professionally qualified status, which is more typically at the equivalent of bachelor's or master's degree levels (Williams and Hanson 2011, 20, 24).

In the UK, the development of higher-level vocational programmes in universities came out of the introduction of work-based learning (defined as learning through, at and for work) in the HE curricula through formal policymaking in the 1980s. Initiatives included accreditation of workplace knowledge and skill through methods associated with the Accreditation of Prior Learning (APL) or the Accreditation of Prior Experiential Learning (APEL) within degrees, 'sandwich' courses involving work experience and vocational foundation degrees (HE intermediate level or Level 5); the latter becoming a recognisable feature of vocational HE in the early noughties onwards (Portwood 2007; Burke et al. 2009). The Leitch Review of Skills in 2006 (HM Treasury 2006) revisited the need to raise minimum levels of adult skills, with particular emphasis on the urgent gap at technician level, which in 2011 was the policy focus of higher apprenticeship development, plus a need to focus on those already in work. Many universities developed provision aimed at people in work matching the four models of good practice for the development of work-based learning programmes identified by Guile and Evans (2010, 15). From a shaky start, where early provision contained little work-based content, foundation degrees emerged that, when linked to professional standards conferred a 'licence to practise' (particularly in the public sector), became the focus of highly innovative work-based provision designed to meet employer needs and targeted by HEFCE funding. It became a key 'selling point' for both learners and employers and an incentive for the involvement of professional bodies as set out in a report on foundation degrees carried out for the Professional Associations Research Network (PARN) that reported, in most cases, foundation degrees were relevant to their sector (Williams and Hanson 2010, 17). This relevance was repeated when further PARN research recorded that 70 per cent of the professional bodies surveyed considered higher apprenticeships to be a suitable programme to becoming fully qualified and professionally recognised and 73 per cent were interested in developing higher apprenticeships as a progression route (Williams and Hanson 2011, 20, 24). However, in the context of higher apprenticeship development in 2011, the specification recognised a Level 5 foundation degree as the highest university qualification for delivery, whereas the report also highlighted attainment at Level 6 as the level of status and skill for professions recognised by most professional bodies and observed:

> This may indicate a need for Apprenticeship frameworks to be expanded to even higher levels to ensure a clear vocational pathway through Apprenticeship to professional status. (Williams and Hanson 2010, 16)

The Wilson Review, with its statements on the need for stronger links between universities and employers, recognised that apprenticeships at the higher level have importance in meeting the "long term skills need of employers" and provided "a highly valued alternative for school leavers who wish to combine work with gaining a higher qualification" (Wilson 2012, 46). The Review also recommended that work-based pathways to HE qualification such as higher apprenticeships should become "a priority development". Perversely, the Review also announced that higher apprenticeships, along with professional qualifications, "are not congruent with the requirements of an honours degree but are equal in rigour and esteem" (Wilson 2012, 46).

The statement that positioned higher apprenticeships as 'not congruent' but 'equal' effectively obfuscated their status and, conceivably, at the time consigned higher apprenticeships to the margins of 'core' or 'mainstream' university business. Positioning was not assisted by a statutory constraint given the highest university qualification that could deliver a higher apprenticeship was at Level 5 (equivalent to a foundation degree or diploma of HE). The advent of Levels 6 and 7 higher apprenticeships that allows inclusion of an honours degree would challenge the oft-repeated mantra that apprenticeships represent an 'alternative' to university, an idea which highlights the struggle amongst policymakers, higher and further education, businesses, learners and agencies responsible for promoting work-based learning to articulate the relationship between practice-based discipline and work-based knowledge and to accept a growing educational debate in favour of how to facilitate professional and technical learning within HE. A subject which this chapter comes back to later.

While clear that 'skills' was not a central concern of university business, progress was being made to engage HE with the apprenticeship agenda with the alignment of apprenticeship frameworks with HE qualifications in the revision to the statutory instrument known as SASE in 2013 following secondary legislation. Significant changes introduced new criteria to recognise apprenticeships at Levels 6 and 7 for the very first time although it is perhaps relevant that some professions including accountancy and law had already established 'apprenticeship-type' routes to qualifying that included degrees—though without very much integration between the degree and working practice. Comprised of at least 120 credits (equivalent to one full-time year of undergraduate study), this new recognition sat alongside revised criteria for provision of at least 90 credits at Levels 4 and 5 and requirements for alignment with professional body recognition (where applicable). Most crucially, SASE allowed for the inclusion of a single integrated qualification that was not reliant on assessing 'knowledge' and 'competence' separately (Bravenboer and Lester 2016, 16). With considerable pedagogic and ideological distance still to travel, these regulatory changes offered a tantalising opportunity for realignment of higher apprenticeships with HE and the prospect of constructive collaboration between employers, professional, statutory and regulatory bodies (PSRB) and universities (Bravenboer 2019, 67). The development of this new type of HE linked to previous work-based learning initiatives, sub-degree provision like foundation degrees and government-funded employability schemes was set to champion workplace learning to deliver higher-level programmes more directly to employers and would become the driver for and reflector of new curriculum development (Lester et al. 2016, 5; Nottingham 2017, 134).

A Secondary Standpoint

From a present-day standpoint, apprenticeships have become a fast-growing and recognisable feature of the higher and professional education landscape. Degree apprenticeships, launched in England in 2015 by the then UK prime minister, David Cameron, were about:

> Equipping people with the skills they need to get on in life and backing businesses to create jobs are key parts of our long-term economic plan. Degree Apprenticeships will give people a great head start, combining a full degree with the real practical skills gained in work and the financial security of a regular pay packet. They will bring the world of business and the world of education closer together and let us build the high-level technical skills needed for the jobs of the future. I want to see many more businesses and universities begin to offer them. (DBIS 2015)

From the outset, degree apprenticeships were intended to bring together the worlds of work and higher education under a dual policy objective: to attract a new cohort of skilled worker with their employer to grow the economy by increasing productivity and to raise levels of social mobility in terms of supporting greater access to HE. The introduction of degree apprenticeships is described as one of the biggest changes to impact on HE in recent decades. Hailed by Jeffrey (2016, 1) as "the greatest opportunity ever seen for anyone concerned with skills and employment" and described by the City and Guilds Group Strategy Skills Board in 2015 as offering a viable 'alternative' to traditional degrees. Crucially, the term 'alternative' in the context of degree apprenticeships becomes laden with meaning.

Intended as complete training programmes designed from the ground up (Edge 2017) higher and degree apprenticeships are described as bringing together and integrating theoretical and practical learning (Lillis 2018; Bravenboer 2019). Positioning them in this way gives the HE qualification a critical role in apprenticeship delivery as higher and degree apprenticeships effectively adopt teaching, learning and assessment methods that are distinct from those of either classroom teaching or more typical vocational training programmes at lower skill levels. They are less formal, independent, practice- or project oriented and collaborative (Lester et al. 2016; Lillis 2018). Such a work-integrated learning (or learning-integrated work) approach is intended to be performative: developing experienced workers' professional practice and effecting workplace change. The latter emphasises transformatory aspects of teaching and learning on the job such as "to think and work outside the box" (Rowe et al. 2017, 187) and incorporating work-based pedagogies and increased individual activities such as "action learning, critical reflection, investigation and to what has been termed practice as research" (Lester et al. 2016, 23). Most importantly, higher and degree apprenticeships reposition our understanding of what is implied and required in developing and accrediting higher-level occupational and professional competence in and through work (Lester et al. 2016; Rowe et al. 2016; Rowe, 2018).

The Issue

Ever since their launch in 2015, degree apprenticeships (and to an extent higher apprenticeships) have been criticised for not delivering on the social mobility agenda. In 2018, at a *very* early stage of their introduction the then skills minister told a House of Lords inquiry into the economics of further, higher and technical education that fears of a 'middle-class grab' on apprenticeships are 'valid' (Offord 2018), and this narrative has continued with higher apprenticeships considered on the one hand a "really great opportunity to boost social justice and social mobility", but "only if we make sure that we widen access to them" on the other (Evans 2022). Indeed, alongside this charge, reports suggest degree apprenticeships are harder to get into than Oxbridge, while perversely 64 per cent of teachers say they would not recommend them to high-achieving students and yet are endorsed by an important constituency: both parents and students (Filby 2023).

So is the criticism fair?

Reflection Needed

Higher and degree apprenticeships have and continue to be a fast-growing feature of HE and skills in England. The value of degree and higher-level apprenticeships has gained considerable traction in alignment over the past decade evidenced by their bringing together the worlds of work and higher education together under dual policy objectives—productivity and social mobility:

> In an advanced economy such as the UK there will be increasing demand for higher-level skills. The ONS (2009) have found that the largest positive contributor to productivity growth has been workers with higher level qualifications. Research from Universities UK (2019) found that the majority (89 per cent) of employers agreed that degrees make apprentices future ready, equipping them with transferable knowledge and 'soft skills' which are vital to remain agile, adaptable, and competitive in a rapidly transforming economy. Higher and degree apprenticeships also support social mobility by enabling many people who previously missed out on higher education to access it. Progression to university in and through work, on an apprenticeship, where previously it had not been seen as an option for the apprentice (or by their parents/carers) is a strong reflection of social mobility impact, whatever the policy context. (Anderson et al. 2022)

There are now those ranked among the foremost universities in the world engaged in apprenticeship delivery (Crawford-Lee and Wall 2018, 239; Crawford-Lee 2020a). It is this aspiration and reality that has provided the challenging and dynamic environment within which the debate about the 'middle-class land grab' is explored. It speaks to what, in essence, an apprenticeship is today.

There is no doubt that it was reassuring when the former skills minister Michelle Donelan promoted degree apprenticeships as a key way to deliver on levelling up and sent a clear signal about how important degree apprenticeships are to the government's agenda. She asked the OfS to rewrite Access and Participation Plans to "strongly encourage provider to set themselves ambitious measurable targets" to significantly increase the proportion of students on higher and degree apprenticeships. In response to the chair of the education select committee Robert Halfon, MP, that the Department for Education (DfE) should introduce a target of 50 per cent of students on degree apprenticeships, the then minister also indicated that she was exploring financial incentives for universities to offer more degree apprenticeship opportunities leading to the secretary of state announcing a Strategic Priorities Grant Fund (SPGF) of £48m. Not since the HAF and the Degree Apprenticeship Development Fund (DADF) has there been such a clear indication that universities needed to get on board in expanding opportunities through higher and degree apprenticeships.

Despite some commentators reporting that "the choice is not between giving people skills or a university education. Universities are Britain's principal purveyors of skills" (Duncan 2023) strong opposition to higher and degree apprenticeships, delivered by universities, has not disappeared from public discourse. Fresh challenges have emerged on issues of quality in delivery, expansion, affordability, use of the apprenticeship levy and value for money (Bewick 2022; Evans 2022; Murphy and Jones 2021), and after all the talk of 'middle-class land-grabs' dominating the discourse in 2021, it has returned. There has been a resurgence of the view that the middle classes are 'hogging' prized apprenticeships (Woolcock 2022) and, indeed, the Sutton Trust, the social mobility charity, published a research report (Cavaglia et al. 2020) that headlined how a greater share of apprentices than ever are from middle-class areas, with a call for reform.

This report has some important messages regarding young people from more deprived backgrounds and the need for more to be done to support them to access degree apprenticeships. However, it misses a critical point, which is that higher and degree apprenticeships have a critical role to play in supporting social mobility and more needs to be done to ensure that individuals from *ALL* backgrounds can benefit from such provision. It is far better to consider social mobility in the context of other key apprenticeship policy objectives, enhancing workforce diversity and inclusion, increasing productivity, supporting the delivery of high-quality public-sector services and delivering the net-zero, green jobs agenda rather than being overly focused on young people. A diverse, well-skilled workforce is a productive workforce.

From a policy perspective, government is clear that apprenticeships are an all-age, all-stage, all-level programme. The impact of apprenticeship on social mobility should, accordingly, be measured on how social mobility is supported for individual, whatever the age. More fundamentally, it could be argued that apprenticeship can make the biggest positive impact on social mobility when it is supported and used as a programme for all ages. This is because apprenticeships and vocational programmes are very different from academic programmes. With traditional academic programmes, progression is mostly linear. Individuals complete GCSEs, progress to A levels (typically) and then progress to fulltime HE at the age of 18 or 19. Vocational and apprenticeship progression is very different. Individuals may leave school or college with an Applied General qualification, work for several years and start a family. In their mid-20s, they may have the opportunity to study and train for a profession through, for example, a degree apprenticeship.

Westminster government understands progression through apprenticeships and vocational programmes and the importance of flexible learning throughout life, as demonstrated by the forthcoming introduction of the Lifelong Learning Entitlement (LLE). If we are to maximise the contribution of higher and degree apprenticeships to social mobility, we must consider and support their use by individuals of all ages. It does not have to be a bifurcated matter. Certainly, more needs to be done to support 18- and 19-year-olds to use degree apprenticeships but equally more need to be done to champion degree apprenticeships that support social mobility and routes into the professions for older individuals. A focus on examples of where and how apprenticeships, particularly degree apprenticeships, are being and could be used, to open new progression routes for underserved individuals and to support workforce diversity, is equally valid. The police constable degree apprenticeship (PCDA) is a particularly good example.

Health and adult social care represent another sector where apprenticeships are being used to develop new opportunities for existing employees, as demonstrated by the growth of nursing associate higher apprenticeship and the potential of the registered nurse degree apprenticeship. This is where new apprenticeship progression routes are supporting social mobility in the National Health Service (NHS) in England.

The failure to fully understand what higher and degree apprenticeships are and how they can support social mobility for individuals of all ages is conflated with a misguided view that they are an alternative to university and therefore more of these opportunities open to, and targeted at, 17-, 18- and 19-year-olds leaving school need creating. Yet, degree apprenticeships are not an alternative to university; they are a university programme typically delivered by a university and involving the award of a degree. Even amongst the ages highlighted, to focus exclusively on school leavers will disadvantage young people of the same ages studying at college or on work-based learning programmes. What is needed is an inclusive approach. School must never be used as shorthand for schools, colleges and independent training providers.

Unstoppable Growth?

In many commentaries, much has been made of the alleged misuse of apprenticeships to train senior leaders and managers. Indeed, it is easy to get the impression that management apprenticeships are a very bad thing: that there has been an 'unstoppable' growth in their use, that they represent a misuse of the apprenticeship levy, a land grab by the middle-class 'haves', and every GBP spent on management apprenticeships is a pound less to spend on 16–19 apprenticeships. Look at the evidence, and the validity (or not) of such claim is apparent. Apprenticeships are a skills programme and poor management skills are regarded by most economists as the key determinant explaining the UK's low productivity. As the Institute for Apprenticeships and Technical Education (IfATE) has explained, use of management apprenticeships is entirely appropriate (Mersinoglu 2020). By their nature, management apprenticeships will frequently be taken by older and existing employees. Those questioning whether this is appropriate need to consider their use by the sector making the greatest use of management apprenticeships—the NHS—which is also by far the largest employer contributor to the apprenticeship levy in England. Those advocating for the focus of apprenticeship to be on 'school leavers' must decide whether social good is best delivered by enabling the NHS to use its levy payments to improve management skills and thereby deliver enhanced patient care. Or if the NHS levy payments should be used to fund apprenticeships for young people in other sectors and job roles.

An area of agreement is that the apprenticeship reforms and apprenticeship levy have led to quality substituting quantity (Cavaglia et al. 2020, 35). Higher and degree apprenticeships are an aspirational programme. For generations apprenticeships and vocational education were regarded as second best. Does the skills system in England want to return to the position where apprenticeships are seen as 'a good choice for other people's children'? It is critical that apprenticeships are *not* just seen as a social mobility programme. Instead, as government intends, apprenticeship should be positioned as a high-quality aspirational programme for everyone.

So, while the Cavaglia et al. (2022) rightly highlights the need to create more opportunities for young people, such an objective must not be delivered in isolation and at the expense of supporting apprenticeships to deliver their other policy objectives: enhancing productivity, supporting inclusion and diversity in the workforce, improving public sector service delivery, supporting the net-zero agenda and of course enhancing social mobility.

Government's "intention to get universities to offer more degree apprenticeships and higher technical qualifications and to help under-served young people" and adults in work realise their potential (Crawford-Lee 2020a). Higher and degree apprenticeships totally meet the objective that higher education focuses on "getting on rather than just getting in" (Donelan 2021). Indeed, the former minister of state for higher and further education, supported greater numbers of higher and degree apprenticeships, more universities and HEIs delivering and more individuals progressing into HE, as a key aim (Crawford-Lee 2020b).

Opposition of higher and degree apprenticeships on the grounds that such programmes are expensive, divert funding from young people who need lower-level entry job programmes, provide 'cost free' access to 'off-the-shelf' and 'rebadged' graduate schemes, are abused by employers (e.g. Dawe 2019; Gravatt 2019) and that openly support a level of discourse that they are not proper apprenticeships at all, but 'fake apprenticeships' or jobs 'masquerading as apprenticeships' (Richmond 2020; Richmond and Eagon 2022) mean that there will be an ongoing debate over the fundamental purpose of the apprenticeship system in England, and the success and future of higher and degree apprenticeship policy (Crawford-Lee 2022).

Not Considering Social Mobility in Isolation

Nobody can underplay the importance of the social mobility agenda but it must be set alongside and considered in the context of the other policy objectives of degree apprenticeships. More broadly, the success of higher and degree apprenticeships needs to be assessed in terms of how they contribute to raising productivity and the delivery of high-quality public sector services. Indeed, they are providing opportunities for social mobility for individuals throughout their working lives.

If an individual will have, on average, five to seven careers during their lifetime, they need to develop a base from which they can acquire such skills. Of course, English, maths and digital skills are of fundamental importance, but so too are other skills that will support an individual acquire new and update existing skills throughout life. In apprenticeships for young people, there is a need to ensure that there is an increasing focus on skills that support and facilitate learning throughout life, and not just on the skills that individuals need to do a job today.

An over-focus on the young and lost opportunities for underserved cohorts will not help apprenticeships deliver social mobility. Individuals should have the ability to progress to a higher-level occupation and fulfil their potential throughout their working life. Unlike traditional academic study, with apprenticeships progress is typically not linear. Individuals do not complete a Level 3 apprenticeship at 18 or 19 and immediately start a higher or degree apprenticeship. What higher and degree apprenticeships can do is provide new opportunities for older adults to train for and secure higher technical, professional and managerial roles. Think of a 30-year-old healthcare assistant from a deprived background having the opportunity to train as a nursing associate or a 35-year-old nursing associate training as a registered nurse. Such opportunities need to be developed and championed. Measures of social mobility that simply focus on achievements by the age of 24 (often associated with the argument of young people with social capital benefitting from opportunities the most) undermine and undervalue the achievements of such individuals.

Those who insist on vocalising the idea of a middle-class land grab deny higher and degree apprenticeships the recognition for helping transform the image and perception of apprenticeships from the good choice of 'other people's children' or the choice for those who could not go to university. They also ignore how the degree 'professionalises' a job role, helps attract a different talent supply and raises performance standards for the occupation. Fundamentally, there is a lack of recognition that occupational 'competence' can be incorporated into university qualifications with degree apprenticeships leading to greater integration between academic qualifications and professional recognition, often allowing the two to be awarded together. The introduction of degree apprenticeships and the power of the degree brand in England have helped sell the benefits of the wider family of apprenticeships—including those at lower levels—with the opportunities for progression in education and a career now seen as integral to each other (Crawford-Lee and Moorwood 2019).

So, while apprenticeships are of course of significant value for many young people and their use should be supported and championed, so too should apprenticeships for older workers. In its 2020 Vision for Apprenticeships, Westminster government was clear that apprenticeships should be an attractive offer for young people and adults. To be of maximum benefit, apprenticeship must remain and be championed as an all-age, all-stage, all-level programme.

Celebrating Success

We need to remind ourselves that apprenticeships in England are a success story. While the apprenticeship system is not perfect, apprenticeships as reformed should not be considered, as the CBI has described previously, a 'failed experiment'. The integrity of the apprenticeship brand should not be damaged by such claims. Instead, apprenticeships are increasingly being seen as an aspirational programme in no small part *because* of higher and degree apprenticeships. Apprenticeships have moved from being an intermediary and provider-led programme with little focus on skills gaps and shortages to a high-quality employer-led programme where apprenticeships are focused on the real skills needs of employers and the UK economy.

Indeed, under the government reforms apprenticeships have been developed and are being used to train individuals in key public sector occupations, in nursing, policing, social work, teaching and adult social care. In the private sector, apprenticeships are being used to tackle skills gaps and shortages in engineering and digital occupations. The proportion of lower-level apprenticeships in questionable occupations has significantly declined. When combined, the apprenticeship levy, the apprenticeship reforms and the introduction of degree apprenticeships have arguably made more impact on achieving academic and vocational parity of esteem than any other education policy in the last 50 years. Today, apprenticeships are approved for delivery for nearly 700 occupations and are increasingly a route into the registered professions. A Degree Apprenticeship represents 'the best of both worlds', the combination of academic and vocational learning, a degree and an apprenticeship. It is because apprenticeships have become such an aspirational programme that concerns over a perceived middle-class grab of apprenticeship places at Levels 6 and 7 are raised at all. Going forward apprenticeships must be developed and delivered as a programme appropriate for everyone, regardless of background. Apprenticeships must never again be seen as the programme 'for other people's children' or provision primarily for those from disadvantaged backgrounds and those failed by the school system.

The next chapter explores the idea of social mobility further and its importance to the purpose of technical education in England. Indeed, social mobility objectives should be determined in the context of what technical education is designed to do. As higher and degree apprenticeships are and will be increasingly attractive to the middle classes (this is not a bad development), technical and professional educational pathways should develop as an attractive option for those of all backgrounds too.

References

Anderson, A., and D. Hemsworth. 2005. *From Apprenticeship to Higher Education: A Guide to Developing Work-Based Progression Routes to Higher Education for Advanced Apprentices and other Skilled Staff and Managers.* Bolton: UVAC.

Anderson, A., D. Bravenboer, and D. Hemsworth. 2012. The Role of Universities in Higher Apprenticeship Development. *Higher Education Skills and Work-based Learning* 2 (3): 240–255.

Augar, P. 2019. *Review of Post-18 Education and Funding.* London: Stationery Office, (CPP 117).

Bathmaker, A.M. 2003. The Expansion of Higher Education: A Consideration of Control, Funding and Quality. In *Education Studies. Essential Issues*, ed. S. Bartlett and D. Burton, 169–189. London: Sage.

Bawden, T. 2011. Degree or Apprenticeship: Which Works Best in the Job Market?, *Guardian* [online] 15 May. Available at: https://www.theguardian.com/money/2011/may/15/graduate-jobs-university-work-experience-apprenticeships (Accessed 31 January 2023).

Bewick, T. 2022. *Boosting Degree Apprenticeships, Supplementary Submission.* House of Commons Education Select Committee, 12 July.

Bravenboer, D.W. 2009. *Commodification and the Official Discourse of Higher Education.* London: University of London, Institute of Education.

Bravenboer, D. 2016. Why Co-design and Delivery Is a "no-brainer" for Higher and Degree Apprenticeship Policy. *Higher Education, Skills and Work-Based Learning* 6 (4): 384–400.

———. 2019. The Creative Disruption of Degree Apprenticeships in the UK. In *Global Perspectives on Work-based Learning Initiatives*, ed. J. Talbot, 57–83. IGI Global: Hershey, PA.

Bravenboer, D., and S. Lester. 2016. Towards an Integrated Approach to the Recognition of Professional Competence and Academic Learning. *Education + Training* 58 (4): 409–421.

Burke, L., D.J. Marks-Maran, A. Ooms, M. Webb, and D. Cooper. 2009. Towards a Pedagogy of Work-based Learning: Perceptions of Work-based Learning in Foundation Degrees. *Journal of Vocational Education and Training* 61 (1): 15–34.

Cable, V. 2010. A New Era for Universities. *Oral Statement to Parliament* [online] 15 July. Available at: https://www.gov.uk/government/speeches/a-new-era-for-universities (Accessed 31 December 2022).

———. 2011. Universities UK Annual Conference. *Oral Statement to Parliament* [online] 8 September. Available at: https://www.gov.uk/government/speeches/universities-uk-annual-conference%2D%2D2 (Accessed 31 December 2022).

Carter, J. 2009. *Progression from Vocational and Applied Learning to Higher Education in England: A Report Prepared for the Rt. Hon. David Lammy MP, Minister of State for Higher Education*. Bolton: UVAC.

Cavaglia, C., S. McNally, and G. Ventura. 2020. Do Apprenticeships Pay? Evidence for England. *Oxford Bulletin of Economics and Statistics* 82 (5): 1094–1134.

Cavaglia, C., McNally, S. and Ventura, Guglielmo. 2022. The Recent Evolution of Apprenticeships: Apprenticeship pathways and participation since 2015. Centre for Vocational Education Research; The Sutton Trust: London

Coughlan, S. 2011. Graduates—The New Measure of Power. *BBC* [online] 2 March. Available at: https://www.bbc.co.uk/news/business-12597811 (Accessed 30 January 2024).

Crawford-Lee, M. 2012. 'Championing higher apprenticeships', *Higher Education, Skills and Work-based Learning*, 2(3), pp. 285–287.

———. 2019. Why Apprenticeships Are More Than a Path to Social Inclusion for the Young. *FE Week* [online] 25 May. Available at: https://feweek.co.uk/2019/05/25/why-apprenticeships-are-more-than-a-path-to-social-inclusion-for-the-young/ (Accessed 13 November 2023).

———. 2020a. Reflecting on a Ministerial Ambition to Widen Participation in HE Through Technical Education and Apprenticeships. *Forum for Access and Continuing Education*, online. Available at: https://www.face.ac.uk/blog-post/reflecting-on-a-ministerial-ambition-to-widen-participation-in-he-through-technical-education-and-apprenticeships/ (Accessed 19 September 2023).

———. 2020b. *The Value of the Degree in Leadership and Management Apprenticeships: a Literature Review*. Bolton, UK: UVAC. Available at: https://uvac.ac.uk/uvac-literature-review-in-defence-of-the-degree/ (Accessed 15 August 2023).

———. 2022. The Lifelong Loan Entitlement, Apprenticeship Levy and Employer Funding Training—Policy Contradictions and Practical Implications. *FE News*, 5 May. Available at: https://www.fenews.co.uk/exclusive/the-lifelong-loan-entitlement-apprenticeship-levy-and-employer-funded-training-policy-contradictions-and-practical-implications/ (Accessed 14 August 2023).

Crawford-Lee, M., and S. Moorwood. 2019. Degree Apprenticeships: Delivering Quality and Social Mobility. *Higher Education, Skills and Work-based Learning* 9 (2): 134–140.

Crawford-Lee, M., and T. Wall. 2018. Sustainability 2030: A Policy Perspective from the University Vocational Awards Council. *Higher Education, Skills and Work-based Learning* 8 (3): 233–242.

Dawe, M. 2019. Apprenticeship Funding: Hard Choices, Even Harder Decisions. In *Future Proofing Apprenticeship Funding in England for the 2020s*, ed. M. Lemin and J. Wright, 22–23. Newcastle-upon-Tyne: NCFE/Campaign for Learning.

Department for Business, Innovation and Skills (DBIS). 2009a. *Specification of Apprenticeship Standards for England*. London: Stationery Office.
———. 2009b. *Higher Ambitions: The Future of Universities in a Knowledge Economy*. London: Stationery Office. Available at: https://www.ed.ac.uk/files/atoms/files/bis-higherambitions-summary-nov2009.pdf (Accessed 1 February 2023).
———. 2010. *Skills for Sustainable Growth: Strategy Document*. London: DBIS. Available at: https://assets.publishing.service.gov.uk/government/uploads/system/uploads/attachment_data/file/32368/10-1274-skills-for-sustainable-growth-strategy.pdf (Accessed 1 February 2023).
———. 2011. *Higher Apprenticeship Fund*. London: Stationery Office. Available at: https://www.gov.uk/government/news/19-000-new-higher-apprenticeships-to-deliver-skills-for-growth (Accessed 10 February 2023).
———. 2015. *English Apprenticeships: Our 2020 Vision*. London: Stationery Office. Available at: https://assets.publishing.service.gov.uk/government/uploads/system/uploads/attachment_data/file/482754/BIS-15-604-english-apprenticeships-our-2020-vision.pdf (Accessed 16 August 2023).
———. 2016. *FE Data Library: Apprenticeships*. Available at: https://data.gov.uk/dataset/fd31be67-82c1-475d-a16f-0b6ddff9f5c4/fe-data-library-apprenticeships (Accessed 24 August 2023).
Department for Education and Employment (DfEE). 1998. *Higher Education for the 21st Century—Response to the Dearing Report*. London: HMSO.
Department for Education and Skills (DfES). 2003. *The Future of Higher Education (White Paper)*. London: HMSO.
Donelan, M. 2021. *This Government Will Ensure Universities Deliver Real Social Mobility*, *Conservativehome*, 24 November. Available at: https://conservativehome.com/2021/11/24/michelle-donelan-this-government-will-ensure-universities-deliver-real-social-mobility/ (Accessed 14 August 2023).
Duncan, E. 2023. It's Ex-polys That Are Showing Unis the Future. *The Times*, 20 October 2023. Available online at: https://www.thetimes.co.uk/article/its-ex-polys-that-are-showing-unis-the-future-p0bvhq9rj (Accessed 15 November 2023).
Employment and Training (England) The Apprenticeships (Specification of Apprenticeship Standards for England) Order 2011. SI 2011/2019. Available at: https://www.legislation.gov.uk/uksi/2011/219/pdfs/uksi_20110219_en.pdf (Accessed 3 February 2023).
Evans, S. 2022. *Raising the bar: Increasing employer investment in skills*. Leicester, UK: National Learning and Work Institute.
Filby, E. 2023. 'As the Middle Classes Rush to Degree Apprenticeships, Poorer British Kids Miss Out. *CITY A.M.* Available online at: https://www.cityam.com/as-the-middle-classes-rush-to-degree-apprenticeships-poorer-british-kids-miss-out/ (Accessed 30 October 2023).

Gravatt, J. 2019. Apprenticeship Funding in England: A Busted Programme Budget. In *Future Proofing Apprenticeship Funding in England for the 2020s*, ed. M. Lemin and J. Wright, 9–10. Newcastle-upon-Tyne, UK: NCFE/ Campaign for Learning.

Gribble, C., J. Blackmore, and M. Rahimi. 2015. Challenges to Providing Work Integrated Learning to International Business Students at Australian Universities. *Higher Education, Skills and Work-Based Learning* 5 (4): 401–416.

Guile, D., and K. Evans. 2010. *Putting Knowledge to Work: Re- contextualising Knowledge Through the Design and Implementation of Work-based Learning at Higher Education Levels.* Lichfield: fdf.

Hall, G., Joslin, H., and Ward, J. 2010. Developing Higher Apprenticeships in England. *Lifelong Learning Networks National Forum Policy Paper* [online] May. Available at: https://www.heacademy.ac.uk/system/files/developing_higher_apprenticeships_in_england.pdf (Accessed 29 January 2023).

Helyer, R. 2015. Adapting to Higher Education: Academic Skills. In *The Work-based Learning Students Handbook*, ed. R. Helyer, 2nd ed., 13–30. Basingstoke, UK: Palgrave Macmillan.

HM Treasury. 2006. *Leitch Review of Skills Final Report: Prosperity for all in the global economy - world class skills.* London: HMSO.

———. 2011. *Plan for Growth.* London: HM Treasury and Department for Business Innovation and Skills.

House of Lords Economic Affairs Committee. 2018. *Treating Students Fairly: The Economics of Post-School Education.* London: TSO.

Jeffrey, C. 2016. *Is Your Vision the Same or Similar to Mine—Can You See What I See?* Association of Employers and Learning Providers, Bristol. Available at: www.aelp.org.uk/ news/aelp-blog/ details/is-your-vision-the-same-or-similar-to-mine-can-you/ (Accessed 26 July 2023).

Joslin, H. 2011. *Implementing Higher Apprenticeships—Learning from Practice.* London: University of Greenwich.

Leitch, S. 2006. *Prosperity for all in the Global Economy—World Class Skills.* London: Stationery Office.

Lester, S., D. Bravenboer, and N. Webb. 2016. *Work-integrated Degrees: Context, Engagement, Practice and Quality. A Literature Review for QAA.* Gloucester: Quality Assurance Agency.

Lillis, F. 2018. *Best Practice in Work-Integrated Learning for Degree Apprenticeships.* London: Middlesex University.

McInerney, L. 2018. Degrees for the Rich, Apprenticeships for the Poor—That's Not a World of Parity. *The Guardian*, 18 December. Available at: https://www.theguardian.com/education/2018/dec/18/degrees-rich-apprencticeships-poor-not-parity (Accessed 17 January 2023).

Mersinoglu, Y. C. 2020. Coupland: Public Funding for Management Degree Apprenticeships Is Perfectly Legitimate. *FE Week*, 31 January. Available online at: https://feweek.co.uk/coupland-public-funding-for-management-apprenticeships-is-perfectly-legitimate/ (Accessed 20 December 2023).

Murphy, H., and Jones, E. 2021. *Apprenticeships at Level 4 and above*. Leicester, UK: National Learning and Work Institute (LWI).

National Apprenticeship Service (NAS). 2011. *Higher Apprenticeship Fund Prospectus*. Coventry: National Apprenticeship Service.

Newton O, Laczic A and Percy C. (2017). *Our Plan for Higher Education*. The Edge Foundation.

Nottingham, P.M. 2017. Re-evaluating Work-based Learning Pedagogy. *Higher Education, Skills and Work-Based Learning* 7 (2): 129–140.

Offord, P. 2018. Fears Over "middle-class grab" On Apprenticeships Are Valid, Minister Admits. *FE Week*, 13 March 2018. Available online at: https://feweek.co.uk/fears-over-middle-class-grab-on-apprenticeships-are-valid-minister-admits/ (Accessed 2 October 2023).

Portwood, D. 2007. Towards an Epistemology of Work-based Learning: Eliciting Clues from Work-based Learning Projects. In *Work-based Learning Futures: Proceedings from the Work-based Learning Futures Conference*, ed. D. Young and J. Garnett. UVAC: Bolton.

Richmond, T. 2020. *Runaway Training: Why the Apprenticeship Levy is Broken and How to Fix It*. London: EDSK.

Richmond, T., and E. Eagon. 2022. *No Train, No Gain: An Investigation into the Quality of Apprenticeships in England*. London: EDSK.

Rowe, L. 2018. Managing Degree Apprenticeships Through a Work-based Learning Framework: Opportunities and Challenges. In *Enhancing Employability in Higher Education Through Work-based Learning*, ed. D. Morley, 51–69. Cham: Palgrave Macmillan.

Rowe, L., D. Perrin, and T. Wall. 2016. The Chartered Manger Degree Apprenticeship: Trials and Tribulations. *Higher Education, Skills and Work-based Learning* 6 (4): 357–369.

Rowe, L., Moss, D., Moore, N. and Perrin, D. 2017. 'The challenges of managing degree apprentices in the workplace: a manager's perspective'. *Journal of Work-Applied Management, 9*(2), pp. 185–199.

Shepherd, J. 2012. Russell Group Extends Membership to Four More Universities. *Guardian*, 12 March. Available at: https://www.theguardian.com/education/2012/mar/12/russell-group-four-more-universities (Accessed 22 February 2023).

Trow, M. 2007. Reflections on the Transition from Elite to Mass to Universal Access: Forms and Phases of Higher Education in Modern Societies since WWII. In *International Handbook of Higher Education*, ed. J.J.F. Forest and P.G. Altback, 243–280. Dordrecht: Springer.

UKCES. 2010a. *Skills for Jobs: Today and Tomorrow, The National Strategic Skills Audit for England 2010, Volume 1 (Key Findings)*. London: UKCES.

———. 2010b. *Skills for Jobs: Today and Tomorrow, The National Strategic Skills Audit for England 2010, Volume 2 (Evidence Report)*. London: UKCES.

———. 2010c. Progression from Vocational and Applied Learning to Higher Education across the UK: A Comparative Study by the University Vocational Awards Council [online]. Available at: https://dera.ioe.ac.uk/1299/1/424657_progression%20to%20higher%20Ed.pdf (Accessed 13 April 2023).

Wall, T., and D. Perrin. 2015. *Slavoj Žižek: A Žižekian Gaze at Education*. London: Springer.

Williams, C., and W. Hanson. 2010. *Foundation Degrees: Barriers and Opportunities in the Professional Bodies Sector*. Bristol: PARN.

———. 2011. *Higher Apprenticeships and Professional Bodies*. Bristol: PARN.

Wilson, T. 2012. *A Review of Business-University Collaboration*. London: DBIS.

Wolf, A. 2011. *Review of Vocational Education—The Wolf Report*. London: Department for Education.

Woolcock, N. 2022. Middle Classes Stand Accused of Hogging Prized Apprenticeships. *The Times*, 8 December. Available online at: https://www.thetimes.co.uk/article/apprenticeships-vs-university-degrees-uk-elite-competition-middle-class-rthjcxtwc (Accessed 17 January 2023).

———. 2023. Starmer Hints at Shift on Tuition Fees. *The Times*, 23 August. Available online at: https://www.thetimes.co.uk/article/keir-starmer-university-too-expensive-for-me-now-bfgrv2gb6 (Accessed 1 November 2023).

CHAPTER 6

Social Mobility and Technical Education: Comfortable Bedfellows?

John Baldwin

Introduction

This chapter is concerned with equity in relation to technical skills education and training. Does everybody have access to this education or training, whether they are young or old, or where they live, their social class, whatever their abilities or gender? This is important for them but is also important for the country and the economy if the abilities of everybody are not being used to their full extent.

Social Mobility and Education

A key indicator of equality is the extent to which the population of a country has high or low social mobility. The Social Mobility Commission (2023a) defines social mobility as

> the link between a person's occupation or income and the occupation or income of their parents. Where there is a strong link, there is a lower level of

J. Baldwin (✉)
Student services, Stamford College, Stamford, UK

J. Baldwin et al. (eds.), *The Future of Technical Education*,
https://doi.org/10.1007/978-3-031-70756-8_6

social mobility. Where there is a weak link, there is a higher level of social mobility.

The UK ranks poorly compared with other developed nations. In 2018, it ranked 12th out of 14 with only the US and Switzerland performing worse (Goldman Sachs 2022). The Scandinavian countries and Canada ranked highest based on social mobility measured by intergenerational elasticity of income. This measure compares the income of children with their parents to see if the children's income is higher. In the UK this difference is small. There are a number of factors that determine whether a person has a higher income and opportunities than their parents, but education can play a major part (The Equality Trust 2023).

The impact is felt at all levels of education. Little et al. (2023) found that children who are taught well in their first year at primary school go on to achieve better grades in GCSE English and Maths. Similarly, Cattan et al. (2014) found that attending preschool could have a significant effect on GCSE outcomes when a pupil reaches 16. Research by Hodge et al. (2021) looked at the difference that one grade makes at GCSE. Overall, they found that on average the impact of one grade made a difference in total lifetime earnings of £23,000. However, the figure is higher for men than for women and is different depending on the subject—the return on maths is twice as high as it is for English. Those who are not eligible for free school meals get a better return that those who do. Education and good education make a difference.

Similar principles apply to those who progress from GCSEs to A levels or other Level 3 qualifications at further education (FE) level. The value added by taking a Level 3 qualification again is higher for men than for women, but men get a better return if they study engineering, while women get a better return over their lifetime if they study business administration or law (The Social Mobility Commission 2023b). The Social Mobility Commission (2023b) found similar difference for those who progressed from Level 3 to higher education (HE) Levels 4, 5 and 6, although they found higher-level return for men at Level 4 and for women at Level 5. A first-degree grade makes a difference; women earn 3.5 per cent and men 7 per cent more if they achieve a first-class degree compared with those who achieve a 2.1. In HE, technical and vocational subjects such as engineering, maths, economics and law give a better financial return that other subjects. The Commission also found that the choice of university makes a difference with graduates from academically selective universities gaining a greater premium, but these universities are less accessible to those from lower socio-economic backgrounds. However, the Russell Group universities claim that the accessibility of those from lower socio-economic backgrounds to their universities is gradually improving (Le Carpentier and Chandler 2023).

This chapter explores these variations in gender, geographical or socio-economic background but also the impact of age or disability on individuals' access to the levels of education—including academic education but focusing on technical and vocational education. However, it will also examine the impact that the Covid-19 epidemic had on education and whether collaboration between education providers at all levels will improve equality and widen participation to technical education. This will include research carried out by the writers.

Widening Participation into Further and Higher Education

Geographical/Socio-economic Deprivation

As we have seen, the level of education a person attains can have a significant effect on their earnings throughout their lifetime. Most of the research has concerned widening participation into higher education, and substantial amounts of money are expended by universities and colleges in persuading young people to progress to HE. However, it is not only earnings that are on average greater for those who progress to HE. Studies in the US (Hathaway 2020) and in the UK (Head 2021) have demonstrated that life expectancy is lower for those who do not progress into further and higher education. The UK study demonstrated that life expectancy is lower in certain communities in the UK where education attainment is low and unemployment and poverty are high, and, in those communities, life expectancy declined during the period before the Covid-19 outbreak. A report for the Department for Health and Social Care (2021), found that adults who have undertaken tertiary education are half as likely to be obese, are 16 per cent less likely to smoke and have fewer co-morbidities. The World Bank (2021) suggests that further and higher education improve chances of getting an interesting job and improve prospects but also leads to higher civic responsibility. Further and higher education benefits individuals but also their communities.

However, these benefits only accrue to those who are able to take advantage of and are motivated to progress on to further or higher education. There is very little research on the barriers to progressing onto Level 3 study either academic via A levels, or vocational qualifications. The government has recognised the need for adult learners to improve their qualifications and prospects and for those over 24 have enabled them to take some Level 3 qualifications which have been identified as having strong wage outcomes and meet key skill priorities free of charge (Jones et al. 2023). However, in their report Jones et al. (2023) identified other barriers for adults to return to education at Level 3. These included general living costs because to return to study adults needed to reduce their working hours or give up working all together, the cost of course materials and travel and childcare costs. In their research, they discovered that 44 per cent of their respondents were not aware of Learner Support or the Advanced Learner Loan Bursary Fund.

As described above, achieving good grades at GCSE at the age of 16 has an impact on the wages that a young person will achieve later in their life. However, achieving good GCSEs and achieving good grades in GCSE English and Maths is critical in being able to access Level 3 education in schools and colleges. Machin et al. (2020) examined the impact of not achieving a grade C in GCSE English had on a young person's future. (When they were researching, grades Cs were required to progress to Level 3. GCSEs are now graded by number, and the equivalent of grade C is now generally Grade 4.) They calculated that achieving a grade D rather than a C decreased the chance of taking A levels by 9 per centage points and increased the chances of dropping out of school by 2 per centage points. They also found that the difference between achieving a grade D in GCSE English rather than C is essentially random in the small window between the two grades. As we have seen above, those from a poorer socio-economic geographical background can easily be disadvantaged by half a GCSE grade in each of their GCSE examinations (Adams and Weale 2022). On behalf of the Sutton Trust, Holt-White and Cullinane (2023) examined the progress of a group of high academic potential young people during the period from the end of primary school and taking GCSEs. They found that of those from an advantaged background, 63 per cent went on to complete A levels, 65 per cent went to university and 24 per cent attended the higher tariff, such as Russell Group universities. In comparison, they identified that of those who suffered disadvantage during secondary schooling (based on eligibility for free school meals), only 45 per cent completed A levels, 53 per cent attended university and 15 per cent attended Russell Group universities.

Crenna-Jennings (2018) identified that by Key Stage 4 (the latter years of primary education), disadvantaged young people were 18.4 months behind in their education compared with their advantaged peers. For her study, she used the Department for Education (DfE) definition of disadvantage which is those pupils who are eligible for free school meals. There are other measures of disadvantage and in various combinations including access to social and cultural capital. Geographical area disadvantage was analysed by Herr (2022), where the top GCSE grades in science subjects including maths were compared. In those counties with low disadvantage such as Berkshire, Buckinghamshire, Surrey and Rutland, over 50 per cent of young people achieved the highest grades, while in counties such as Bedfordshire, Lincolnshire and Suffolk, where deprivation is higher, only about 33 per cent achieve the high grades. This in turn impacts on young people from poorer backgrounds being able to progress onto A levels or Level 3 vocational science courses and ultimately progress on to science courses at university.

In Explore Education Statistics, Gov UK (2023) reported that in 2021/2022 the results gap for Level 3 qualifications, A levels, applied general and applied technical qualifications between disadvantaged and non-disadvantaged students was the highest since disadvantage comparison began to be made in 2016/2017. The average grade for advantaged students was B at A level and for disadvantaged was C+. Similar gaps were identified for those taking Level 3 vocational and technical qualifications such as BTEC qualifications. This gap can of course be the difference between being accepted on to a course at university or not. The two main qualifications for entry to higher education are A levels and BTEC Level 3 on their own or in combination. Students who take BTEC qualifications only are more likely to come from lower socio-economic backgrounds (Dilnot et al. 2022). BTEC students are more likely to drop out of university and get a lower overall degree than those who enter with A levels. Kelly (2017) found that almost all the students from lower socio-economic backgrounds who had entered HE since 2008 had taken BTEC qualifications.

Similar points are made by Lisauskaite et al. (2021), who emphasise that a higher proportion of young people who have been entitled to free school meals attend further education colleges and thus are much more likely to take vocational/technical qualifications than those who attend sixth forms in schools. They suggest that this in turn affects progression to higher education, as those who have taken vocational/technical qualifications at 16–18 are, again, much more likely to take similar courses in HE. However, Lisauskaite et al. (2021) found in their research that those from a disadvantaged background who attended an FE college are more likely to ultimately achieve an HE qualification in a vocational/technical subject than those from a similar background who attended a school sixth form. Those disadvantaged young people who attended an FE college are less likely to apply to a Russell Group university but those who do attend, achieve as well as those who are more advantaged. The writers conclude by suggesting that if the country needs a higher-trained technical workforce, then FE colleges can probably contribute more than they do currently.

One of the major contributors to technical education are apprenticeships. Yet, the number of apprenticeships taken in 2022/2023 was 4.6 per cent down from the previous year, although the overall number of higher apprenticeships taken is rising. In Chap. 5, it is discussed whether those from deprived backgrounds are less likely to be accepted for an apprenticeship than their better-off peers. The Social Mobility Committee (2020) suggests that this is generally the case and that those from poorer backgrounds tend to be accepted for lower-level apprenticeships. The Sutton Trust and UCAS (Sutton Trust 2023) report that most degree apprenticeships are increasingly being utilised by large levy-paying companies for professional development for experienced staff, mainly in London, and thus young people from disadvantaged backgrounds are being crowded out of degree apprenticeships opportunities. UCAS include an apprenticeship research facility on their website (UCAS 2023).

Both higher and degree apprenticeships are similar, and both involve working while studying. Higher apprenticeships refer to apprenticeships that include the achievement of academic/vocational qualification from Level 4 up to bachelor's and master's degree at Levels 6–7. All levels can include both vocational qualifications and academic qualifications. However, Degree Apprenticeships involve achieving a full bachelor's or master's degree (Levels 6 and 7) as a core component of the apprenticeship (Pearson 2023).

Gender

In 2020/2021, the progression rate to university for females was 50.6 per cent compared with 38.4 per cent of males (Dept of Education 2022d), and 56 per cent of undergraduates were female and 44 per cent male (Advance HE 2022). The Higher Education Statistics Agency (HESA) produces an annual report on Who is Studying in HE, and they distinguish between science subjects and non-science subjects. In 2020/2021, 701,460 female students studied a science subject compared with 594,385 male students. However, not all science subjects could be described as technical, and in subjects such as engineering and technology there were 37,965 female students compared with 147,455 male students (HESA 2023).

A report produced by science, technology, engineering and maths (STEM) learning concluded that at GCSE level there is little difference between male and female students accessing science subjects. However, female students tended to achieve high passes in biology, while male students tended to achieve high passes in physics, and, consequently, at A level there is a higher proportion of male students taking physics and female students taking biology (STEM Learning 2022). Similarly, over 14 per cent of male students take A level maths compared with 9 per cent of female students. When gender and disadvantage are examined together, the differences are starker. About 4 per cent of female students who have been in receipt of free school meals (FSMs) are entered for A level biology compared with about 15 per cent of female students who have not been in receipt of FSMs. Similarly, with A level physics, about 4 per cent of male students who have been in receipt of FSMs are entered compared with 10 per cent of male non-FSM students. McDool and Morris (2022), have demonstrated that these gender differences also exist within STEM vocational education. They suggest that the gap is even wider if the young person is taking a vocational qualification at both Level 2 and Level 3. They believe that this is substantially driven by gender stereotypes. Women are more likely to take vocational qualifications in hair and beauty and in social care, while men are more likely to take construction and engineering qualifications including apprenticeships in these areas. They also suggest that female students are more risk-averse than male students and thus will only take a subject such as maths, physics and engineering if they believe that with their prior experience they are likely to do well. McDool and Morris (2022) also discovered that when taking into account socio-economic status (SES), a girl from a lower SES is more likely to take a vocational STEM qualification than an A level STEM qualification, whereas for boys their SES does not appear to have the same effect. In addition, they noted that women who take a STEM subject and particularly if they progress on to higher education are ultimately likely to have a higher income than men, whereas if they opt for a 'traditional' female occupation such as care or hairdressing, their income is likely to be lower than for men who opt for engineering, plumbing or to be an electrician.

Ethnicity

The level of qualification young people attain overall differs for different ethnicities. Table 6.1 shows the average GCSE Attainment 8 score by ethnicity. (This is based on averaging the points for the 8 GCSEs that pupils take but counting English and maths twice.) It also shows the average points score for A levels, Applied General and T level qualifications by ethnicity. (The points score is based on specified points for grades received.)

As can be seen, in Table 6.1, at GCSE level both Asian and Chinese students achieve higher grades, but on taking A levels, Chinese students still score higher but Asian students fall below the overall average. For Applied General qualifications such as BTEC Extended Diplomas, the difference is less stark, although Chinese students get slightly higher grades. However, for T levels the gap between Chinese and all other ethnicities is wider. For all three types of qualification, black students perform poorer than students from other ethnicities.

Table 6.1 The average GCSE Attainment 8 score by ethnicity

Ethnicity	*GCSE Attainment 8 score*	*A levels—average points score*	*Applied General average points score*	*T levels—general average points score*
All	51	42	33	32
Asian	56	40	33	32
Black	50	37	30	30
Chinese	69	46	34	38
Mixed	51	41	31	30
White	50	41	33	32

Source: DfE (2022, 2022a)

Table 6.2 Eighteen-year-old entry rate into higher education by ethnic group 2021

Ethnicity	*Percentage entry*
Asian	55
Black	49
Chinese	72
Mixed	41
White	33

Source: DfE (2022b)

As would be expected, young people from different ethnicities are more or less likely to progress to higher education and thus have the opportunity of achieving a degree-level qualification, as illustrated in Table 6.2.

The entry rate for white students is lower because of a lower average overall application rate, not a lower acceptance rate. Table 6.3 gives an approximation of the types of subjects that students from different ethnicities are more likely to take (Bolton and Lewis 2023).

As can be seen, Asian and black students are more likely to take technical subjects than mixed or white students. Connor et al. (2004) suggest that these differences arise primarily because often parents of non-white young people are keener for their children to take professional/vocational routes which they believe will give them higher incomes and stability. This begins with choices at GCSE and continues at A level.

Age

In 2019/2020, 37 per cent of all undergraduates at UK universities were mature students aged over 21 years. However, most of them were taking Level 4/5 or part-time courses. Moreover, 28 per cent of these mature students were studying part-time compared to 3 per cent of young people and 79 per cent of students on Level 4/5 technical courses were mature students. Mature students are more likely to be female, have a disability and come from a lower socio-economic background (Hubble and Bolton 2021a). The Office for Students reported that universities have not prioritised the recruitment of mature students even though the number entering HE, particularly on a part-time basis, has fallen. The Office indicated that mature students play a very important part in upskilling the workforce, which is needed for economic reasons. Thus, universities should investigate how they can meet the needs of mature students (Office for Students 2020).

Table 6.3 What subjects were different groups of students more or less likely to study? Full-time home students starting undergraduate courses in 2022

Ethnicity	*Most likely to study*	*Least likely to study*
Asian	Medicine and dentistry, computing, law, engineering/ technology, maths	Agriculture/food studies, veterinary science, geography/environment, design/ arts
Black	Subjects allied to medicine, computing, business/ management	Veterinary science, agriculture/food studies, geography/environment, history/ philosophy/religion
Mixed	Combined/general studies, languages, design/arts	Education, agriculture/food studies, subjects allied to medicine
White	Agriculture/food studies, veterinary science, geography/ environment	Medicine and dentistry, computing, law, subjects allied to medicine, engineering/ technology

Source: DfE (2022b)

In the UK, the Open University recruits the highest number of part-time distance learning students with 68 per cent over 25 years of age (The Open University 2023). In 2020/2021, just over 40,000 of its students were taking technical subjects and over 46,000 taking social science and arts subjects (The Open University 2023). In that year, 63 per cent of its students were female, 88 per cent were white and 25 per cent of its students lived in the 25 per cent most deprived areas. Over 40 per cent of its students declared a disability.

In the UK, there is an assumption that higher education study is mainly for young people, and they go away to university at the age of 17/18 to take a three-year undergraduate degree. The government introduced the Lifelong Loan Entitlement (LLE) in order that people can take HE courses that meet the needs of employers. The entitlement will come into effect from 2025 and will be for £37,000 tuition fees, which can be used on a flexible basis at universities or colleges to take short courses, modules or full HE courses. The aim is to encourage people to access HE whenever they need it during their lifetime (Department for Education 2023). One aspect of the entitlement is that modules and short courses, as well as traditional degree courses, will be priced consistently according to the amount of learning they contain. This should create a fair, more flexible system to help support and encourage more people into post-18 education. It is also designed to encourage more people to take Levels 4 and 5 courses including Higher Technical Qualifications. It is suggested that people will be able to take modules over a long period of time building up to a full Higher Technical Diploma or a full degree (DfE 2023a).

The entitlement to a loan however will only apply up to the age of 60 (DfE 2023a). Susan Pember, who is the policy director of HOLEX, which represents adult education providers, suggests that this is ageist, and as the retirement age will be 67 by 2025, this age limit is unwise as people may still wish to retrain at age 60 or above (Chowen 2023). Similarly, Crawford-Lee (2023) argues that apprenticeships are not just for the young, although she suggests that may in the skills sector argue that the young should be prioritised. She points out that giving a 30-year-old the opportunity to take a nursing apprenticeship might be better value for the nation than for a 17-year-old to take a hairdressing apprenticeship where there may be less of a skills shortage.

However, the director of the Higher Education Policy Institute suggests in a report that there will be a number of stumbling blocks to making the entitlement successful (Stephenson 2023). She argues that there will be difficulty in accumulating different modules from different colleges and universities with administrative problems as to who 'owns' the student for reporting reasons. She also suggests that most students will use up most of their entitlement when they take a full-time three-year degree. In the report, it is pointed out that the entitlement is for taking 30 to 40 credit modules but for people who are trying to upskill they would often prefer smaller modules. Only modules contained in current qualifications can be accessed which will limit choice.

However, the entitlement could contribute to both levelling up and meeting the needs of those from less-advantaged backgrounds who might want to study at a slower rate and/or by distance learning, and as the first programmes to be included are Higher Technical qualifications, this should meet the need for more technical rather than academic qualifications. There are a large number of Levels 4 and 5 qualifications validated by universities and colleges and by qualification bodies such as Pearson Edexcel. However, in order to encourage more people to take this level of qualification, the government introduced Higher Technical Qualifications. These are subject to an approval process by the employer-led Institute for Apprenticeships and Technical Education to ensure they meet a quality mark, thus giving both students and employers the reassurance that the qualifications are of a high standard (The Institute for Apprenticeships and Technical Education 2023). The first programmes were launched in 2022.

Disability

The disabilities that people have will all impact on their ability to study and access education but to a varying extent. The Special Educational Needs and Disability Act 2001 gave disabled people rights not to be discriminated against in education and training. The definition of disability is broad and includes both physical and mental disabilities including specific learning difficulties. In 2019/2020, 17.3 per cent of all home university students declared that they had a disability of some kind (Hubble and Bolton 2021). In 2018/2019, 17.8 per cent of young people with special educational needs (SEN) had started higher education compared with 47.3 per cent of those without a SEN. In 2026/2017, of those taking an apprenticeship, 9.8 per cent declared a learning difficulty (TVET Inclusion 2018). In a report, the British Council (2018) points out that providing effective TVET could improve a disabled person's income by between 55 and 95 per cent and thus reducing taxpayer costs by about £170,000 during their lifetime. Moreover, there would also be direct economic benefits to local authorities by reducing adult social care costs. Prior to 2020, students with disabilities had tended to have lower retention and achievement rates than those without a disability, but in 2020–2021 students with disabilities were reported as having continuation rate of 0.2 per cent higher than those without disabilities. Similarly, in 2021–2022 qualifiers who reported a disability were 0.7 per cent higher than those who did not report a disability (Office for Students 2023b). It could be suggested that these improvements are as a result of universities becoming more skilled at supporting students with disabilities.

Kernohan (2019) analysed HESA data and discovered that the highest proportion of HE students with disabilities take subjects such as fine art, ceramics, combined social studies, Latin, archaeology, drama and photography. Thus, comparatively few take technical subjects.

However, some people with disabilities, such as those with autism, have particular strengths which enable them to perform jobs requiring technical skills. Cope and Remington (2022) suggest that people who are autistic have strengths which include pattern recognition, logical and systematic problem-solving, attention to detail due to enhanced local processing and sustained focus and can be innovative and creative because they see the world differently.

Intersectionality

We have seen above that a person's socio-economic background, sex, age, ethnicity and whether they are disabled or not can determine whether they are likely to progress to higher education. However, some individuals are likely to fall into a number of these categories. A particular focus has been on white, working-class boys. An Education Committee report in 2021 found that only 16 per cent of white boys who were eligible for FSMs progressed to HE by the age of 19 (UK Parliament 2021). This compared with FSM-entitled Chinese students with a 72.8 per cent participation rate and FSM-entitled black African pupils with a 59 per cent participation rate. Raven (2022) found that cost concerns, financial and opportunity, are a greater deterrent for advancement to HE by white working-class boys in North-West England than for their more affluent peers. However, Benson-Eggleton (2021) suggests that the emphasis on working-class boys hides the fact that white working-class girls only do slightly better.

Funding for Widening Participation to HE

The funding for encouraging those from under-represented background to progress on to higher education comes from a number of sources but the main source is from the universities and colleges themselves. Almost all universities and colleges are required by the Office for Students to produce an Access and Participation Plan (APP) in which they agree to set aside money to improve access to HE but also to support those from under-represented backgrounds when they are undertaking HE (Office for Students 2023a). Crawford-Lee (2022) argues that in its requirements it should be encouraging HE providers to actively include degree apprenticeships and technical education in their APPs. She suggests that the focus of most APPs is on raising awareness of HE generally and aspirations, whereas HE providers ought to be signposting progression pathways to technical professional status, particularly to those from under-represented backgrounds. This would provide motivation to those who are undecided or wary of accessing HE and a clear understanding of what HE study could lead to.

In its guidance, the Office for Students (2023b, 24) does suggest that HE providers should have intervention strategies that include working with employers and with Uni connect partnerships to promote degree apprenticeships.

So, What Does All This Mean for Technical Education?

Above, we have seen that there are categories of people who are less likely to progress to higher education but are there also categories of people who are less likely to progress on to particular academic subjects in higher education? Some evidence has been provided to suggest that women are less likely to progress on to some technical HE courses and that some autistic people are more likely to do so. Are people from deprived areas and backgrounds more or less likely to progress into technical subjects in HE? The Royal Economic Society commissioned a study into why certain categories of people do not progress onto studying economics, arguably an academic subject compared with business studies which is more applied (Fuentes et al. 2023). They found that economics can be seen as an elitist discipline at university with the lowest proportion of students from under-represented backgrounds compared with all other disciplines. This lack of social diversity was most pronounced in Russell Group universities.

BTEC courses are designed to be vocational/technical, and Gicheva and Petrie (2018) found that in 2015/2016 54 per cent of students applying to university were taking A levels, 18 per cent were taking BTEC qualifications and 8 per cent were taking a combination of both. In 2016, 37 per cent of black students entered university having taken a vocational qualification compared with a national average of 24 per cent for all ethnicities; 32 per cent came from a working-class background, and 47 per cent of young people from the lowest participation into HE areas entered with a vocational qualification.

Thus, overall, young people from a working-class background whose parents are less-well-off who apply to university are more likely to have taken a vocational/technical qualification than A levels compared to the general population.

The Impact of Covid-19

The Covid-19 pandemic had a profound effect on education worldwide. Howard et al. (2021) reported that in England during the spring and summer terms 2020, schools and colleges were not very effective in providing learning concentrating on the use of worksheets and videos. In the autumn term of 2020, schools reopened but teaching was disrupted as individual students or students in 'bubbles' were affected by Covid and had to self-isolate, and thus teachers often had to provide face-to-face teaching and online at the same time. In the spring of 2021, England went into lockdown again, although teachers had become more adept at providing effective online learning. By March 2021, most students returned to school or college, although there were still some restrictions, but, for most pupils, learning returned;

Thus, the learning of almost all students was adversely affected by the pandemic but there were particular concerns about those from low-income families, particularly when online learning was prevalent and many of these students had little or no access to digital tablets or laptops. The Committee of Public Accounts (2023) heard evidence that disadvantaged children were the most badly affected during the pandemic, wiping out ten years of progress in reducing the gap in attainment between them and their peers. They also were concerned that it was predicted that it would take another decade for disadvantaged students' attainment to return to the position it was before Covid. Schlindwein et al. (2023) interviewed 17-/18-year-olds who had been affected by the Covid pandemic and who were 'high attainers'. They found that those from disadvantaged backgrounds felt that they perceived differences in the poorer quality of teaching they received and the poorer access they had to digital devices, enabling them to take part in on-line learning.

Atherton (2020) carried out an international comparison of support in higher education for those from deprived backgrounds during the pandemic and found that different countries responded in different ways. Most of the 45 countries surveyed at least provided some support to purchase laptops except England and Japan, which did not. Moreover, most countries provided some relief from tuition fees for those from poorer backgrounds including Scotland, Wales and Northern Ireland, but England did not.

A form of learning that became noticeable during the Covid pandemic was place-based education. Yemini et al. (2023) explained that place-based learning emphasises the connection between the learning process and the place or area where the learning is taking place and the use of different learning environments such as working from home or teaching and learning outdoors. This style of learning could be said to apply to technical education and the use of work-based or work experience where the learner experiences learning on the job in a local business. The expression, place-based education has also been used as part of the levelling up agenda for links to be made between universities, FE colleges and employers in a locality (Civic University Network 2022).

Levelling Up

The Levelling Up White Paper set out how the government will spread opportunity more equally across the UK (Department for Levelling Up, Housing and Communities 2022). It indicates that it is concerned with geographic differences and social mobility and one of its policies is to significantly increase the number of people successfully completing high-quality skills training in every area of the UK. As part of the levelling up agenda, the government promised to invest in ensuring that 200,000 more people successfully completing high-quality skills training annually, the creation of nine new institutes of technology and the creation of a Future Skills Unit, which will provide better analysis of employer technical skills needs (Patel 2022). In a report, the million-plus group of universities (a group of 22 modern universities) suggested that the boosting of the role of institutes of technology will help more young people achieve both T levels and higher technical qualifications at Levels 4 and 5 with both universities and FE colleges involved (Million Plus 2020). Similarly, Blond and Ostrowski (2022) argue that there should be an expansion in further education colleges offering Level 5 and five qualifications, as these qualifications are often more focused on employer's practical needs and colleges offer a more local place to study on a full- or part-time basis. These could meet the needs of those young or older people who are reluctant to go away to university to take a traditional three-year programme. This could help to boost the economy in those levelling up regions which need regeneration and technical employment skills to help drive that regeneration and where there is a low level of HE participation.

However, Navelles (2022) argues that the expansion of T levels will have a negative effect on levelling up because the qualifications require students to undertake the equivalent of two days a week on work experience and in rural communities, such as in Devon and Cornwall, where there are communities with high levels of deprivation, that lack industry, placements will not be available. Similarly, FE colleges may not be able to offer these narrow programmes, in sparsely populated areas because there will be insufficient demand to make viable class sizes.

Pope et al. (2022) examine how successful they believe levelling up will be in regenerating regions of the UK. They confirm that increasing the number of young people taking higher education qualifications in an area can be successful, particularly in technical subjects, although they believe that Levels 4 and 5 qualifications and higher apprenticeships are equally important as three-year degrees. However, they point out that unless the jobs are available in a region, graduates will not remain. They believe that lifelong learning is very important but also point out that the people who take qualifications later in life tend to be those who already have higher qualifications. Moreover, if there are no jobs in an area that require higher qualifications or training, then local people will not be motivated to reskill. Similarly, they emphasise that productivity will remain low in an area if people are over-skilled and frustrated; thus, there must not be a mismatch of skills.

Collaboration

A point that is made repeatedly is that technical skills training in England falls behind many countries and even Scotland, Wales and Northern Ireland because of the 'missing middle' education and training at Levels 4 and 5. The Augar Review (DfE 2019) found that despite the need for technical education at Levels 4 and 5, the number of enrolments were falling and enrolments on STEM higher technician courses were a small minority. They found that for universities there is a financial incentive to provide three-year degree programmes, and although the cost of technical courses is higher than for social science courses, the fees they receive are the same. The review emphasised that for those who did not want to take a three-year course or those who were reluctant to commit to three years of tuition loans Levels 4 and 5 courses are a valuable alternative, but few are available at universities. The Skills and Post 16 Education Act (DfE 2022c) implemented some of the Augar proposals and is intended to improve access to technical qualifications and make it easier to access tuition loans on a flexible basis over a student's lifetime.

The Lifelong Education Institute (2022) referred to the highly successful model already in operation in Scotland, with articulation agreements between further education colleges and universities which enable students to spend the first one or two years of higher education in an FE college, either by taking a Higher National Certificate or Higher National Diploma or by completing a two-year foundation degree. They then finish their training on a '2+2' year basis with a two-year 'top-up' bachelor's degree at a university. Such agreements exist in England between many further education colleges and universities but have been less successful for a number of reasons. This is also discussed in Chap. 3.

Bhattacharya and Norman (2021) carried out a review of competition and collaboration of universities and further education colleges in the UK. They suggested that there are many reasons why colleges and universities should work together. Firstly, they believe that universities can validate HE courses to be delivered in colleges where there is no local university, thus allowing less-mobile students to study at HE level along with those who may feel more comfortable in a smaller college environment. Secondly, articulation agreements between a university and college will allow less-advantaged students to be able to take a Level 4 or 5 course at a college and then progress on to a degree course at the university. Thirdly, by working together they can better meet local skill needs because by pooling resources and expertise, they can offer programmes jointly that they could not deliver on their own, and colleges often have closer links with local employers. Fourthly, they can share knowledge as colleges have more experience in working with widening participation students, whereas universities can bring their research knowledge. Finally, they can share resources and facilities if they are in the same city or town. Colleges often have better basic engineering facilities than universities. An example is the Teesside University College Partnership, where the university and local colleges have combined in order to support regeneration of the Redcar steelworks. Bhattacharya and Norman (2021) also make the point that colleges recruit a much higher proportion of disadvantaged students than schools, which often feel more comfortable taking HE courses in FE colleges, particularly technical courses. They also found that although there are many advantages to collaboration, there are also barriers as often universities and colleges are competing for the same students. By working together, more people from under-represented backgrounds can access higher education while also increasing the number of people taking higher technical qualifications.

The Shape of the Future Labour Market and Conclusion

It is plainly difficult to predict how the labour market will change in the future and the types of skills and knowledge will be required. However, Universities UK (2023) predict that by 2035 the number of graduates needed will increase from 16 million to 26 million and that by 2035 88 per cent of new jobs will be for graduates. They predict that a high proportion of these jobs will be technical jobs in computing or engineering. As we have seen, the last few years there have been a range of government initiatives including the introduction of the Lifelong Loan Entitlement, Higher Technical Qualifications, T levels and an increased promotion of apprenticeships all of whom have been described above. What will the future hold for technical education and particularly for those who are currently disadvantaged?

Potentially, the Lifelong Loan Entitlement, which is now rebranded as the Lifelong Learning Entitlement, will have the greatest impact for disadvantaged students who wish to study a technical subject. They will not have to take the traditional three-year degree programme at the age of 18, but they and others will take HE modules as they need them in their career. However, this initiative, which has been described above, will mean a substantial change in the way that colleges and universities will deliver courses. The chief executive of Universities UK acknowledged that this was going to be a challenge, particularly for universities but will also be a great opportunity that will require collaboration between universities and colleges (Stern 2023). The Department for Education has carried out an assessment of the feasibility of the LLE, and the permanent secretary to the department has acknowledged that the economic appraisal has demonstrated that the LLE programme is likely to deliver value for money. She said that it is expected that the entitlement will incentivise more people to study at Levels 4 to 6, including those who would not have otherwise studied because of the barriers they face in the current system (Patel 2023). However, she indicated that this is a very complex operation which will involve many government departments as well as universities and colleges, and she warned that the timescale of implementation in 2025 may not be achievable.

A major part of this initiative is the flexibility which will enable people to study individual modules over a period of time at different institutions. As discussed above, Stephenson (2023) explains that this is fraught with difficulties because most degree modules are 30 credits but can sometimes be 20 or 10 credits to meet particular needs. Could a person start off studying modules in construction and then switch to modules in marketing, and will every university be content to accept other universities modules, and who will award the final qualification? The Office for Students is consulting on this. The government assumes that this will all be in place by 2025. She points out that to make this flexibility work, course modules will have to be offered in an evening and at weekend. The course fee will probably be fixed in line with current rules; so, will providing such course be viable for colleges and universities? However, as indicated above, higher technical qualifications, which can be provided by further educational colleges, could be more flexible. FE colleges have experience in offering part-time courses and are often more accessible for disadvantaged learners who wish to take up study later in life. However, they sometimes have difficulty in recruiting qualified staff in technical subjects because of the rates of pay they can offer compared with universities.

The undertaking of apprenticeships from Level 1 to Level 6 should continue to provide the technical knowledge and skills that the country needs in a very wide range of occupations, but it is more difficult to predict that T levels will be of benefit to many disadvantaged learners, as discussed in Chap. 2. It is suggested by the DfE that the introduction of the Advanced British Standard will ensure that academic and technical education at secondary level will be equal to young people taking a combination of five subjects with the potential mix of major and minor academic and technical subjects (DfE 2023b). The future of technical education and the labour market will be discussed in much more detail in Chap. 7.

Summary

This chapter has continued the theme of technical education by sharing research on social mobility, levelling up and the impact of gender, age, disability and social class on the provision of technical education. It has attempted to demonstrate the barriers that many members of the community, who have the potential to contribute, face if they wish to consider or access technical education and how some of those barriers can be overcome. This includes not only collaboration between universities and further education colleges but also, as discussed in Chap. 3, greater collaboration between further education colleges.

References

Adams, R., and S. Weale. 2022. GCSE Grades Gap for Disadvantaged Pupils in England Widest in a Decade. *The Guardian*, 20 October 2022. Available at. https://www.theguardian.com/education/2022/oct/20/gcse-grade-gap-disadvantaged-pupils-england-widest-decade-child-poverty-attainment (Accessed 30 April 2023).

Advance HE. 2022. *Equality in Higher Education: Statistical Reports 2022*. York: Advance HE. Available at https://www.advance-he.ac.uk/news-and-views/equality-higher-education-statistical-reports-2022 (Accessed 17 May 2023).

Atherton, D. 2020. *University Access, Student Success and COVID-19 in a Global Context*. The Sutton Trust. Available at https://www.suttontrust.com/wp-content/uploads/2020/11/Covid-and-Global-University-Access (Accessed 9 July 2023).

Benson-Eggleton, J. 2021. *Gendered Education Policy Does More Harm to White Working-Class Girls Than Discussing White Privilege*. WONKHE. Available at https://wonkhe.com/blogs/gendered-education-policy-does-more-harm-to-white-working-class-girls-than-discussing-white-privilege/ (Accessed 31 October 2023).

Bhattacharya, A., and A. Norman. 2021. *Study Buddies? Competition and Collaboration between Higher Education and Further Education*. London: Social Market Foundation.

Blond, P., and M. Ostrowski. 2022. *The Government Should Make Further Education the Powerhouse of Levelling Up*. WONKHE. Available at https://wonkhe.com/blogs/the-government-should-make-further-education-the-powerhouse-of-levelling-up/ (Accessed 2 July 2023).

Bolton, P., and J. Lewis. 2023. *Equality of Access and Outcomes in Higher Education in England*. House of Commons Library. Available at https://researchbriefings.files.parliament.uk/documents/CBP-9195/CBP-9195.pdf (Accessed 16 June 2023).

British Council. 2018. *Disabled People's Inclusion within UK Technical and Vocational Education and Training*. British Council. Available in https://www.britishcouncil.org/sites/default/files/disabled_peoples_inclusion_within_uk_tvet (Accessed 12 August 2023).

Cattan, S., C. Crawford, and L. Dearden. 2014. *The Economic Effects of Pre-school Education and Quality*. IFS Report 99. Institute for Fiscal Studies. Available in https://ifs.org.uk/publications/economic-effects-pre-school-education-and-quality (Accessed 27 May 2024).

Chowen, S. 2023. Lifelong Loan Entitlement : Maintenance Loans in and ELQ Rule Out. *FE Week*, 12 September 2023. Available in https://feweek.co.uk/lifelong-loan-entitlement-maintenance-loans-in-and-elq-rule-out (Accessed 11 August 2023).

Civic University Network. 2022. *Going Further and Higher: How Collaboration between Colleges and Universities Can Transform Lives and Places*. Civic University Network. Available in https://civicuniversitynetwork.co.uk/fe-he-collaboration/ (Accessed 27 May 2024).

Connor, H., C. Tyers, T. Modood, and J. Hillage. 2004. *Why the Difference? A Closer Look at Higher Education Minority Ethnic Students and Graduates*. London: Department for Education and Skills.

Cope, R., and A. Remington. 2022. The Strengths and Abilities of Autistic People in the Workplace. *Autism in Adulthood* 4 (1): 22–31.
Crawford-Lee, M. 2022. *A Vocational View on Access and Participation Is Needed from the Regulator.* WONKHE. Available in https://wonkhe.com/blogs/a-vocational-view-on-access-and-participation (Accessed 20 August 2023).
———. 2023. *Apprenticeships Are Not Just for the Young*.... University Vocational Awards Council. Available in https://uvac.ac.uk/apprenticeships-are-not-just-for-the-young (Accessed 13 August 2023).
Crenna-Jennings, W. 2018. *Key Drivers of the Disadvantage Gap—Literature Review.* Education Policy Institute. Available in https://epi.org.uk/wp-content/uploads/2018/07/EPI-Annual-Report-2018-Lit-review.pdf (Accessed 1 May 2023).
Department for Education. 2019. *Independent Panel Report to the Review of Post-18 Education and Funding*, (the Augar Report). London: DfE. Available in https://www.gov.uk/government/publications/post-18-review-of-education-and-funding-independent-panel-report (Accessed 27 May 2024).
———. 2022. *Ethnicity Facts and Figures—Average Score for Students Taking A Levels and Other Qualifications.* London: DfE. Available in https://www.ethnicity-facts-figures.service.gov.uk/education-skills-and-training/a-levels-apprenticeships-further-education/average-score-for-students-taking-a-levels-and-other-qualifications/latest/ (Accessed 27 May 2024).
———. 2022a. *Ethnicity Facts and Figures—GCSE results (Attainment 8).* London: DfE. Available in https://www.ethnicity-facts-figures.service.gov.uk/education-skills-and-training/11-to-16-years-old/gcse-results-attainment-8-for-children-aged-14-to-16-key-stage-4/latest (Accessed 3 June 2023).
———. 2022b. *Post-16 Education Outcomes by Ethnicity in England.* London: DfE. Available in https://assets.publishing.service.gov.uk/government/uploads/system/uploads/attachment_data/file/1086958/Post-16_Ethnicity_Topic_Note.pdf (Accessed 4 June 2023).
———. 2022c. *Skills and Post-16 Education Act.* London: DfE. Available in https://www.legislation.gov.uk/ukpga/2022/21#:~:text=An%20Act%20to%20make%20provision,make%20provision%20about%20student%20finance (Accessed 27 May 2024).
———. 2022d. *Widening Participation in Higher Education: 2022.* London: DfE. Available in https://www.gov.uk/government/statistics/widening-participation-in-higher-education-2022 (Accessed 27 May 2024).
———. 2023. *Everything You Need to Know about the Lifelong Learning Bill.* London: DfE. Available in https://educationhub.blog.gov.uk/2023/02/02/everything-you-need-to-know-about-the-lifelong-learning-bill (Accessed 23 July 2023).

———. 2023a. *Lifelong Loan Entitlement, Government Consultation Response*. London: DfE. Available in https://assets.publishing.service.gov.uk/media/64060e398fa8f527f310d54b/Lifelong_Loan_Entitlement_Consultation_Response.pdf (Accessed 27 May 2024).

———. 2023b. *The Advanced British Standard: Everything You Need to Know*. London: DfE. Available in https://educationhub.blog.gov.uk/2023/12/14/the-advanced-british-standard-everything-you-need-to-know (Accessed 16 December 2023).

Department for Levelling Up, Housing and Communities. 2022. *Levelling Up the United Kingdom*. London. Available in https://assets.publishing.service.gov.uk/government/uploads/system/uploads/attachment_data/file/1052708/Levelling_up_the_UK_white_paper.pdf (Accessed 20 June 2023).

Department of Health and Social Care. 2021. *Education and Schooling Health Summary*. London. Available in https://www.gov.uk/government/publications/education-schooling-and-health/education-schooling-and-health-summary (Accessed 13 April 2023).

Dilnot, C., L. Macmillan, and G. Wyness. 2022. *Educational Choices at 16–19 and University Outcomes*. Oxford: Nuffield Foundation.

Fuentes, S., T. Burnett, G. Cagliesi, P. Chaudhury, and D. Hawkes. 2023. *Who Studies Economics? An Analysis of Diversity in the UK Economics Pipeline*. Royal Economic Society. Available in https://res.org.uk/who-studies-economics-an-analysis-of-diversity-in-the-uk-economics-pipeline/ (Accessed 27 May 2024)

Gicheva, N., and Petrie, K. 2018. Vocation, Vocation, Vocation, *The role of vocational routes into higher education*, Social Market Foundation.

Goldman Sachs. 2022. *The Bigger Picture, UK Social Mobility-A Tough Climb*. Goldman Sachs. Available in https://www.gspublishing.com/content/research/en/reports/2022/02/24/c9116edf-3ff1-4ec3-a459-bf9889763b0b.html (Accessed 8 April 2023).

Gov UK. 2023. *Academic Year 2021/22 A Level and Other 16 to 18 Results*. London. Available in https://explore-education-statistics.service.gov.uk/find-statistics/a-level-and-other-16-to-18-results/2021-22 (Accessed 30 April 2023).

Hathaway, B. 2020. Want to Live Longer? Stay in School, Study Suggests. *Yale News*, 20 February. Available in https://news.yale.edu/2020/02/20/want-live-longer-stay-school-study-suggests (Accessed 13 April 2023).

Head, E. 2021. Life Expectancy Declining in Many English Communities Even Before Pandemic. *Imperial College London News*, 12 October 2021. Available in https://www.imperial.ac.uk/news/231119/life-expectancy-declining-many-english-communities (Accessed 13 April 2023).

Herr, O. 2022. *Education: What Role Does Location and Deprivation Play in GCSE Results?* British Science Association. Available in https://www.britishscienceassociation.org/blog/education-gcse-results-location-deprivation (Accessed 30 April 2023).

HESA. 2023. *Who's Studying in HE*. HESA. Available in https://www.hesa.ac.uk/data-and-analysis/students/whos-in-he (Accessed 27 May 2024).

Hodge, L., A. Little, and M. Weldon. 2021. *GCSE Attainment and Lifetime Earnings Research Report*. London: DfE. Available in https://assets.publishing.service.gov.uk/media/60c36f0cd3bf7f4bd11a2326/GCSE_Attainment_and_Lifetime_Earnings_PDF3A.pdf (Accessed 27 May 2024).

Holt-White, E., and C. Cullinane. 2023. *Social Mobility: The Next Generation, Lost Potential at Age 16*. The Sutton Trust: London. Available in https://www.suttontrust.com/wp-content/uploads/2023/06/Social-Mobility-The-Next-Generation-Lost-Potential-Age-16.pdf (Accessed 27 May 2024).

Howard, E., E. Khan, and C. Lockyer. 2021. *Learning During the Pandemic—Review of Research From England*. London: Ofqual. Available in https://www.gov.uk/government/publications/learning-during-the-pandemic/learning-during-the-pandemic-review-of-research-from-england (Accessed 9 July 2023).

Hubble, S., and P. Bolton. 2021. *Support for Disabled Students in Higher Education in England*. House of Commons Library: London. Available in https://commonslibrary.parliament.uk/research-briefings/cbp-8716/ (Accessed 27 May 2024).

———. 2021a. *Mature Higher Education Students in England*. London: House of Commons Library. Available in https://commonslibrary.parliament.uk/research-briefings/cbp-8809/ (Accessed 27 May 2024).

Jones, E., Y White, S. Hall, and M. Bexon. 2023. *Cost Barriers to Learning at Level 3—Research Report*. London: DfE. Available in https://assets.publishing.service.gov.uk/government/uploads/system/uploads/attachment_data/file/1139923/Level_3_Cost_Barriers_Report_March2023.pdf (Accessed 14 April 2023).

Kelly, S. 2017. *Reforming BTECs: Applied General Qualifications as a Route to Higher Education*. Higher Education Policy Institute, Report 94, Oxford. Available in https://www.hepi.ac.uk/wp-content/uploads/2017/02/Hepi_Reforming-BTECs-Report-94-09_02_17-Web.pdf (Accessed 27 May 2024).

Kernohan, D. 2019. *Subjects of Study and Students with Disabilities*. Wonkhe. Available in https://wonkhe.com/blogs/subjects-of-study-and-students-with-disabilities/ (Accessed 19 June 2023).

Le Carpentier, C., and H. Chandler. 2023. *Pathways for Potential 2023*. Russell Group. Available in https://pathwaysforpotential.russellgroup.ac.uk/pathways-for-potential-2023-a-report-by-the-russell-group/index.html (Accessed 27 May 2024).

Lisauskaite, E., S. McIntosh, S. Speckesser, and H. Espinoza. 2021. *Going Further, Further Education Disadvantage and Social Mobility*. London: Centre for Vocational Education Research, and Sutton Trust.

Little, A., L. Hodge, and P. Tymms. 2023. *The Economic Benefits of Effective Reception Classes in England*. London: DfE.

Machin, S., S. McNally, and J. Ruiz-Valenzuela. 2020. Entry Through the Narrow Door: The Costs of Just Failing High Stakes Exams. *Journal of Public Economics* 190: 104–224.

McDool, E., and D. Morris. 2022. Gender Differences in Science, Technology, Engineering and Maths Uptake and Attainment in Post-16 Education. *The Manchester School* 90 (5): 473–499.

Million Plus. 2020. *Levelling Up: Investing in Higher Technical Education at Universities in England*. Available in www.millionplus.ac.uk/policy/reports/levelling-up-investing-in-higher-technical-education-at-universities-in-england (Accessed 25 June 2023).

Navelles, G. 2022. *Levelling Up Skills: Are T Levels Good for Social Mobility?* Business West. Available in https://www.businesswest.co.uk/blog/levelling-up-skills-are-t-levels-good-social-mobility (Accessed 9 July 2023).

Office for Students. 2020. *English Higher Education 2020: The Office for Students Annual Review*. OfS. Available in https://www.officeforstudents.org.uk/annual-review-2020 (Accessed 23 July 2023).

———. 2023a. *Access and Participation Plan Guidance*. Ofs. Available in https://www.officeforstudents.org.uk/advice-and-guidance/promoting-equalopportunities/access-and-participation-plans (Accessed 3 August 2023).

———. 2023b. *Regulatory Advice 6: How to Prepare an Access and Participation Plan*. Ofs. Available in https://www.officeforstudents.org.uk/media/110a2d5e-b6f2-4e2b-b339-628c2a8294ab/regulatory-advice-6-how-to-prepare-an-access-and-participation-plan.pdf (Accessed 4 August 2023).

Patel, N. 2022. *What Would It Mean to Truly Level Up People's Skills Across the UK?* Tech UK: London. Available in https://www.techuk.org/resource/what-would-it-mean-to-truly-level-up-people-s-skills-across-the-uk.html (Accessed 27 May 2024).

Patel, A. 2023. Lifelong Loan Entitlement Faces Significant Timescale Challenges Warns DfE Perm Sec. *FE Week*. https://feweek.co.uk/lifelong-loan-entitlement-faces-significant-challenges-to-meet-2025-launch-warns-dfe-perm-sec/ (Accessed 10 December 2023).

Pearson. 2023. *Apprenticeships 101 A Really Simple Guide to Higher and Degree Apprenticeships*. Pearson. Available in https://qualifications.pearson.com/content/dam/pdf/new-apprenticeships/101Insightfactsheets/C0261c%20factsheet%20AKT13636.pdf (Accessed 27 May 2024).

Pope, T., E. Shearer, and P. Hourston. 2022. *Levelling Up and Skills Policy How Qualifications and Training Can Help Boost Regional Productivity*. Institute for Government: London. Available in https://www.instituteforgovernment.org.uk/sites/default/files/publications/Levelling-up-and-skills-policy.pdf (Accessed 27 May 2024).

Raven, N. 2022. Recognising the Risks: Perceptions of Higher Education Amongst Young White British Males from Areas of Educational Disadvantages. *Research in Post-Compulsory Education* 27 (4): 643–663.

Schlindwein, J.P., M. Warby, M. Holt, H. Kenney, and A.F. Asma. 2023. *Stories from the Class of 2023.* Sutton Trust: London. Available in https://www.suttontrust.com/our-research/stories-from-the-class-of-2023/ (Accessed 28 August 2023).

STEM Learning. 2022. *Gender, Socioeconomic Disadvantage, and Ethnicity Shown to Affect Science Attainment and Progression.* Stem Learning UK, Available in https://www.stem.org.uk/sites/default/files/pages/downloads/Sciencepercent20Educationper cent20inper cent20Englandper cent20Summary.pdf (Accessed 17 May 2023).

Stephenson, R. 2023. *Does the Lifelong Loan Entitlement Meet Its Own Objectives?* HEPI: Oxford. Available in https://www.hepi.ac.uk/2023/05/11/does-the-lifelong-loan-entitlement-meet-its-own-objectives-2/ (Accessed 27 May 2024).

Stern, V. 2023. *The Lifelong Loan Entitlement Is an Opportunity. Let's Grab It.* Universities UK. https://www.universitiesuk.ac.uk/latest/insights-and-analysis/lifelong-loan-entitlement-opportunity (Accessed 10 December 2023).

The Committee of Public Account. 2023. *Education Recovery in Schools in England, Fifty-fifth Report of Session 2022–23.* UK Parliament: London. Available in https://committees.parliament.uk/publications/41567/documents/204875/default/#:~:text=The%20department%20has%20announced%20total,%2F21%20to%202023%2F24 (Accessed 27 May 2024).

The Equality Trust. 2023. *Childhood.* The Equality Trust: Croydon. Available in https://equalitytrust.org.uk/childhood (Accessed 8 April 2023).

The Institute for Apprenticeships and Technical Education. 2023. *Introduction to Higher Technical Qualifications and Scope of Approval.* IfATE: London. Available in https://www.instituteforapprenticeships.org/qualifications/higher-technical-qualifications/introduction-to-higher-technical-qualifications-and-scope-of-approval/ (Accessed 5 August 2023).

The Lifelong Education Institute. 2022. *Skills and Levelling Up Manifesto.* The Lifelong Education Institute. Available in https://www.lifelongeducation.uk/post/skills-and-levelling-up-manifesto (Accessed 15 July 2023).

The Open University. 2023. *Facts and Figures.* OU, Milton Keynes. Available in https://www.open.ac.uk/about/main/strategy-and-policies/facts-and-figures (Accessed 31 July 2023).

The Social Mobility Commission. 2020. *Apprenticeships and Social Mobility: Fulfilling Potential.* Social Mobility Commission: London. Available in https://www.gov.uk/government/publications/apprenticeships-and-social-mobility-fulfilling-potential (Accessed 27 May 2024).

———. 2023a. *About Us.* The Social Mobility Commission: London. Available in https://www.gov.uk/government/organisations/social-mobility-commission/about (Accessed 7 April 2023).

———. 2023b. *Labour Market Value of Higher and Further Education Qualifications: A Summary Report*. DfE. https://www.gov.uk/government/publications/labour-market-value-of-higher-and-further-education-qualifications-a-summary-report/labour-market-value-of-higher-and-further-education-qualifications-a-summary-report (Accessed 13 April 2023).

The Sutton Trust. 2023. *Where Next? What Influences the Choices of Would-Be Apprentices*. Sutton Trust: London. Available in https://www.suttontrust.com/wp-content/uploads/2023/07/Where-next-What-influences-the-choices-of-would-be-apprentices (Accessed 28 August 2023).

The World Bank. 2021. *Education Finance Watch 2021*. The World Bank.

TVET Inclusion. 2018. *Disabled People's Inclusion Within UK Technical and Vocational Education and Training*. British Council: London. Available in https://www.britishcouncil.org/education/skills-employability/research/disabled_peoples_inclusion (Accessed 27 May 2024).

UCAS. 2023. *Apprenticeships*. UCAS: Gloucester. Available in https://www.ucas.com/apprenticeships (Accessed 28 August 2023).

Universities UK. 2023. *Jobs of the Future*. Available In https://www.universitiesuk.ac.uk/what-we-do/policy-and-research/publications/jobs-future (Accessed 27 May 2024).

Yemini, M., Engel, L., and Simon, A. (2023). Place-based Education—A Systematic Review of Literature. *Educational Review*. Available in https://www.tandfonline.com/doi/full/10.1080/00131911.2023.2177260 (Accessed 15 July 2023).

CHAPTER 7

Technical Education in the Next 30 Years

Ross Renton

Background

This chapter explores the future terrain of technical education, employing a methodology that supports future policy formulation. Drawing on insights from the preceding chapters and the opinions of leading policy-makers and sector leaders, the chapter presents a perspective on the future of technical education within the English tertiary education system. Leveraging these insights, the chapter puts forward potential policy levers to facilitate the implementation of future strategies. It concludes with a discussion on how to enhance and communicate these policies, reflecting on the challenges encountered in the past 30 years.

Methodology

The chapter utilises complementary methods, integrating documentary analysis from the previous chapters and a thematic analysis informed by some of the concepts found within futures studies.

R. Renton (✉)
Anglia Ruskin University Peterborough (ARU Peterborough),
Peterborough, UK
e-mail: Ross.renton@aru.ac.uk

J. Baldwin et al. (eds.), *The Future of Technical Education*,
https://doi.org/10.1007/978-3-031-70756-8_7

The documentary research is conducted by examining relevant policy documents and published research, which were subsequently evaluated within the context of the goals of the chapters. The preceding chapters have provided the contribution to the documentary research, offering a breadth of data and enabling the reader to track policy changes across several decades, for example in Chap. 1. The selected methodology is considered a practical approach for analysing technical education policies to support the exploration of potential scenarios for the future trajectory of technical education.

The use of the insights from current policy and system leaders aids the contextual analysis, giving a wide-ranging understanding of the perceived challenges for the future. In addition, a thematic analysis was applied to the insights provided by the diverse range of stakeholders to identify and interpret the themes that have arisen from their observations.

Some of the principles found within futures studies have been employed to support the scenario development. This approach has previously been used by one of the authors within a study on the future relationship between the state and higher education providers, helping to develop future scenarios. Future studies or foresight research has been known by several different names and can have a slightly different meaning when used across various fields (Sardar 2010). Despite these differences, in practice the overarching principles are often similar. A potential future position is shaped by the knowledge of what factors have impacted on the past to forecast which factors might have the most impact in the future.

Futures studies has developed as field in recent years and is now seen as a creditable approach to policymaking, commonly being used across the globe by policymakers and practitioners to help establish decision-making processes, understand potential risk and planning. The UK Civil Service has been using the method for several years, with a specialist unit based within the Government Office for Science. They have developed a range of resources, including a Futures Tool Kit (Government of the United Kingdom 2017).

The Six Pillars approach (Inayatullah 2008) has helped by influencing the approach used within this chapter, assisting in the consideration of the key factors that may impact the future. It has provided a useful framework to consider the future development of policy in relation to technical education without being unnecessarily rigid in the application of the concept.

Expert Insights

A selection of national policy experts, sector leaders and employer representatives were invited to provide their insights to support the development of the future scenarios in this chapter. They were selected based on the insights provided within the earlier chapters and where expertise has been identified as being required to enhance the concepts arising from the chapters. They were given an open-ended question on the future of technical education over the next 30 years, which has given rise to a rich set of qualitive data.

Employers and Industry Representatives

This group of stakeholders highlight the value of technical education within the wider education system. They identify the issue of divides between different forms of education at a system level and the perceived undervaluing of technical education, mirroring some of the points found in Chap. 3. This also raised concern at the individual learner level, where there is a desire to ensure those without traditional qualifications have access to training/education. There are concerns being raised that the current education system is not responsive enough to the needs of employers and that there is a disjoint between providers and industry. This group would like a greater focus on upskilling and reskilling as part of policy reform; however, there is a point made about how disruptive constant reform/change can be for employers. Employers recognise the opportunities that technological advances will have on the workforce and education system; they identify a need for greater digital literacy to support this period of transition.

> It's up to us to build the skills system that we want for the future. If we want a great education and skills system in 30 years' time, we need to start putting in place now a new policy approach to achieve it. First, technical education shouldn't be siloed away but instead integrated as a valuable part of the overall education and skills system, helping everyone get the skills they will need for work and life in the economy for the next 50 years. Secondly, the system must genuinely reflect and respond to labour market needs to avoid having too many people trained in areas with a surplus, while gaps, therefore jobs, remain unfilled. Also, large employers currently dominate qualifications and apprenticeship design, but it's SMEs that hire more – but we have had much less say, so far. Apprenticeship starts at SMEs have almost halved

from 240,970 in 2016/17 to 123,780 in 2020/21, but the right reforms here will be key – diluting the apprenticeship levy to be broader and so large firms can spend the funds but non-levy paying small businesses see less, is not the answer. Nationally and locally, small businesses need to be part of the skills policy process – Federation of Small Business (FSB) has led on a number of Local Skills Improvement Plans (LSIPs); streamlining these would be a good way forward.

Thirdly, upskilling/reskilling needs to be focused on those who'd benefit from it, most. At the moment, the most likely people to upskill later in life are graduates while a large proportion of people leaving school without level 2 or level 3 qualifications are unlikely to train when they're older. Similarly, help for those at work is too focused on salaried employees - access for four million self-employed has been curtailed for new skills if they want to change what they do later in life – a point from FSB that the Chancellor listened to at his Autumn Statement in November 2023, and HMRC will now enact. We must avoid a two-tier education/skills system now emerging; where those who were successful at school end up with all the qualifications. This avoidance of a tiered approach also counts for digital skills and AI - we can't only focus on high level digital skills but lower level and intermediate digital skills are also essential. Fourth and finally, having the next election as the moment to define a clear reform strategy is crucial. We hear so often that constant tinkering in technical education, and more broadly, is a real problem. T-levels and traineeships being introduced then scrapped in a very short time period is a great example to avoid, as manifesto-writers consider the 2025–2030 Parliament and raise their sights to 2050.

(Craig Beaumont, Chief of External Affairs at the Federation of Small Businesses (FSB))

It is not news that many of tomorrow's jobs don't exist today and that people will work for much longer through a 100 year life. But whilst many a report over recent years has discussed how educators and employers could create an improved vocational ecosystem, two pressures are going to force significant change.

Firstly, the UK's demographics are not only changing because we are living longer, in 2022 there were 124,000 fewer births. Our population graph is no longer a pyramid – there are 16 per cent fewer 20–29 year olds than 50–59 year olds. Secondly, if technology was reducing the need for workers, a smaller working population wouldn't be a problem. But the UK needs more people in highly skilled roles, not less, an estimated 2.5m more by 2035.

Employers will change how they recruit, train and retain people. Early talent programmes will focus less on a person's education exit point, more on training and developing people without direct experience – the phrase of the moment for this is skills based hiring. Employers will also focus on internal mobility, retraining and redeploying their existing workforce as buying in talent will become too expensive or difficult. Change in practice will be unavoidable.

(Stephen Isherwood, Joint CEO—Market Intelligence and Insights, Institute of Student Employers)

Whether referencing net zero, de-carbonisation, circular economy, renewables or sustainability, green industry will be a key growth area over the next 30 years that cannot be ignored. In my work areas alone, the UK resources and waste sector has an estimated £10 billion per year turnover and the UK plastic industry represents a further £25 billion per year turnover, combined representing 9000 UK companies and 500,000 UK jobs. Connecting technical education and transferable skills to the requirements of a developing green economy will be vital.

Alignment of long-term education, government policy and industrial development is a must. We need a clear vision for the UK green economy to 2050 and a roadmap by sectors and regions to get there. A fundamental part of this is the need to build and make available relevant apprenticeships, degree courses, on the job learning opportunities and professional materials at scale to realise that transition, enabled by a next generation workforce with the required skills and knowledge. From engineers and scientists to innovators, planners and project managers, a developing green economy will cut across many technical work areas and represents a real opportunity for UK leadership. Studies suggest that the estimated 250k UK jobs in green industries could become upwards of 2 million by 2030 if business and government show the right level of ambition, which will in turn need supporting by the education sector.

(Stuart Foster, RECOUP, Chief Executive Officer. PS Recovery Systems, Managing Director. European Association of Plastic Recovery and Recycling Organisations, Director)

Over the coming decades technology will revolutionise the nature of employment, far faster than ever before and therefore the associated education to fulfill those roles, must also respond with agility. Whilst many employers struggle to articulate the exact nature of their future roles, what is clear is that we will need a workforce who are far more digitally literate and 'work ready'.

Technology may bring the opportunity to train at a lower cost due to the reduced need for capital equipment. This could also enable access for a greater population, allowing disconnected communities to take advantage where investment in technical training facilities, in certain locations may not have been economically viable.

The need to upskill the existing workforce will be essential, providing the technical skills that the economy of the future requires and therefore the funding models will have to adapt. It will be extremely important to support individuals who may earn a lower income or have larger financial commitments such as dependents.

The real test though, will be whether the technology available for utilisation, aligns with the actual processes and equipment being utilised in industry. Therefore, the need for industry and education to work closely remains essential.

(Charlotte Horobin, Chief Executive Officer Cambridgeshire Chamber of Commerce)

Technical education is paramount to the health of UK business; traditional 'academic' education alone does not create a diverse and skilled workforce capable of driving growth and productivity. In the future, it is likely that people will need to be equipped with a core set of technical and transferrable skills, such as green and digital, to navigate a changing world of work dominated by megatrends such as automation and net-zero.

Government, business, as well as education and training providers must all work together to address the barriers preventing people from participating in technical education and progressing onto technical careers. This should start with developing the knowledge infrastructure that can support the identification of new and emerging skills needs and the commitment from all stakeholders to act upon this. This will help ensure technical education is not only an attractive employment route, but a future-focused pathway responsive to the changing forces and features of the labour market.

(Richard Tunnicliffe, Regional Director of the Confederation of British Industry (CBI))

In the next 30 years, technical education will likely witness a deeper integration of artificial intelligence (AI) into curricula, particularly in further education colleges and private training establishments. Challenges may arise in aligning educational programs with the rapidly evolving AI landscape, necessitating close collaboration between manufacturers, government bodies, and educational institutions.

With ongoing concerns about skills shortages government may need to invest in updating infrastructure and providing financial support to ensure that technical education remains relevant and prepares students and professionals for emerging technologies. Manufacturers play a pivotal role in shaping educational content to meet industry demands. They will continue to underpin educational needs by providing product specific technical training. Overall, a collaborative effort among all stakeholders will be crucial for addressing challenges and fostering innovation in technical education. Providing career guidance and development platforms, such as the CIPHE career pathway, will become increasingly important.

Mandatory Continuing Professional Development (CPD) in, safety specific areas of technical education, is likely to become more prevalent in the future. This could involve a shift towards flexible learning models, allowing professionals to engage in ongoing education while balancing work commitments. The integration of technology, such as online platforms and virtual learning environments, will facilitate this flexibility and will enable individuals to keep updated with everchanging learning and educational requirements.

(Kevin Wellman, Chief Executive Officer, The Chartered Institute of Plumbing and Heating Engineering)

The rising importance of technical education will rebalance the landscape of tertiary education over the next few decades. The advent of employer-designed degree-apprenticeships has already been a game-changer. Between 2021 and 2022, LinkedIn job adverts not requiring a traditional academic degree increased by 33 per cent and employers preferring skills qualifications and experience grew by 20 per cent. Elite employers are leading the change, removing the need for 'unnecessary' qualifications from their recruitment lexicon. This is not about dumbing down; it's about adopting a broader view of what great talent looks like, ensuring no talent is wasted, and that everyone, regardless of early life-chances, has the opportunity to succeed.

(Claire Paul, Assistant Director Skills, Cambridgeshire and Peterborough Combined Authority and the former BBC Director of Leadership and Career Development)

Education Sector Representatives

This education sector group focused their responses on lifelong learning, with the need for technical education to be easily accessible through someone's life and career. This echoes some of the points already made in Chap. 5. There is a recognition of the role higher education must play in helping to improve the recognition of technical education and the need for higher education to evolve to meet the future challenges. There are concerns around the perceived artificial barriers between technical and non-technical education shared with employers and this is also a feature of earlier chapters. Innovation within the curriculum and delivery is a key theme, with a desire for greater personalisation and industry-related learning. There is an acknowledgement that the education system needs to exist within a global economy, with a recognition that there is increasing competition for a capable work force and the expansion of higher education in other countries. The respondents want to see greater collaboration with industry to help identify future talent needs and a collective responsibility to meet the needs of the future.

> As I see it, the UK sits at a future-defining crossroads. If the Treasury forecasts are to be believed, we're broke, broken and potentially freefalling towards economic disaster. How we got here will be the subject of much debate and discussion, but for now there needs to be a plan to repair and refocus, reinvent and rebuild for the next 30 years. In every meeting we have with government officials, ministers and shadow ministers, three words abound: productivity, growth, and skills. Education provides the solution to these three challenges. Certainly, our global competitors know this: that's why they are purposely investing in their own higher education systems.
>
> However, I believe the picture will be different in 30 years' time. I believe we will have near universal higher education participation, and the offer will be predominantly technical, vocational and professional in content and delivery. This is attested to time after time by the reports that enter my inbox on a daily basis. Whether it's the World Economic Forum, OECD, ONS or McKinsey, each report forecasts massive unmet demands for high-level skills. These will be in a range of fast-emerging areas: engineering, climate change, healthcare; the challenges of AI, technology, and an increasingly ageing population will be writ large. We don't need a revolution in our higher education system, just evolution and recognition of our key assets. All the building blocks are there, they just need to be nurtured and, frankly, respected. Our professional and technical universities saw us through the industrial revolutions of the past and they will see us through the next ones. Governments of today and tomorrow simply need to see us not as an enemy of the state, but the solution waiting impatiently to deliver.
>
> (Vanessa Wilson, CEO of University Alliance, the higher education group representing leading professional and technical universities across the UK)

As the world around us changes and England's under-18 domestic population rises, one principle should remain constant to guide long-term decision making in education: providing the opportunity for anyone to study who has both the ability and aspiration to do so. It is already clear that the solution to future skills needs and labour market challenges is more education, not less. So, we must work hard to maintain a system that neither puts limits on the number of people who can access tertiary education of all kinds, nor prescribes what disciplines individuals choose to study. While curricula will, and should, develop to incorporate emerging needs, such green skills, digital and artificial intelligence (AI), student choice must remain at the heart of any future education strategy. No government or employer is all-knowing, however well-intentioned, and it is impossible to accurately predict the skills needs of tomorrow.

What is certain, however, is that people are living longer, jobs are changing, and it is more common for individuals to have multiple careers and work in multiple industries throughout their lifetimes. This means the need for lifelong learning is only going to get greater, heightening the need for technical education to be easily accessible to people at any point in their lives and at the level they require it – even if this is at a level lower or equivalent to a qualification they may already hold in another discipline. People's future access to education must not be defined by their pasts, so to keep our education offering future-facing, we need to develop a strategy that is both flexible and fulfilling.

(Dr Diana Beech, Chief Executive Officer, London Higher)

Over the next 3 decades, as we move towards the mid 21st century, technical skills are gradually finding their rightful place centre stage in the British education system. Over the last 100 years, they have played second fiddle to the perceived gold standard of academic learning - largely acquired through a university education.

Technical ability, nurtured in our flagship further education colleges, is now widely recognised by students, parents, politicians and the media as vital to economic, social and personal success. Technical and craft skills have become the 'must have' skills for employment and for building sustainable communities in a interconnected digitally focussed world. Technical education's era is upon us.

(Dame Ann Limb, Chair Lifelong Education Institute)

The artificial boundaries between technical and non-technical will continue to erode. Students across all disciplines will integrate different forms of experiential learning within their formal curriculums. This will be particularly important for students from poorer backgrounds and those undertaking non-residential higher education, who typically face social capital gaps due to not being able to fully participate in the immersive elements of the student experience. By integrating approaches such as service-learning into the curriculum, all students, regardless of background or mode of study will benefit. These students won't be 'work ready' as this requires investment from employers, but they will have all had experiences and skills which enable them to thrive once they leave formal tertiary education.

(Richard Brabner, Executive Chair, UPP Foundation and Director of Environmental, Social and Governance)

Our education systems must shift – and urgently. Preparing technical education students for the future with technology and curricula that will be outdated by the time they finish college or university does little to fix today's skills gaps or create a proficient workforce.

It only takes a little free thinking to envisage what we could achieve. Imagine a world where learning is central to living, not just work. It's skills that matter, not exam results. The learning sparks come from the student after education providers and their industry counterparts have kindled the fire with access, opportunity, resources, information and guidance.

At tertiary level, technical education is personalised, immersive, adaptive and, most importantly, exciting! It allows students to find and use what works for them; it's relevant to real life and gained from different learning situations: pop-up classes, virtual classrooms, AR and VR simulations, collaboration with peers locally and worldwide, and more. Project-based learning and mastery-based progression put an end to exam pressure. There are no institutional boundaries because students are not confined to just one. Instead, they access education from a selection of education providers and industry organisations who actively support them in the areas where they excel.

What students want from their education is already entirely possible with technology; it's our systems that need to change.

(Al Kingsley, Chair Hampton Academies Trust, RSC Advisory Board for the East of England and Author incl. *My School and Multi Academy Trust Growth Guide*)

There has been a great deal of debate on the purpose of education. Is it to give you the skills to do specific jobs or is it something that you can rely on to support you in both your professional and wider personal lives? We also have somewhat arbitrary debates about the similarities and differences between technical, vocational and academic education. In the next 30 years we may wish to let go of the ideological arguments here and instead focus on how the world is changing and what skills we need to thrive. We need a better funding solution to ensure equality of access to education throughout our working lives and we need to value all of the benefits of education, not just the private economic/performance benefits.

We must also get much better at working collaboratively with industry and communicating with each other about work innovations and future skills/workforce needs. Many employers are risk-averse and lack the capacity to engage in long-term solutions, and many HEIs are working on innovative solutions but lack capacity to engage with industry. We need to join these things together. Furthermore, regional skills boards need to be genuinely future thinking, not just what current employers want now or only at sub degree level. It is our collective responsibility to get this right and I hope that over the next 30 years we can create a bottom-up solution that improves equality of opportunities and meets our productivity/skill needs.

(Dr Kate Wicklow, Policy Director (Insight, Access & Student Experience), GuildHE)

Higher education is expanding worldwide, driven by the ambitions of young people and their families, who have identified higher levels of learning as the best means to navigate uncertain futures. Globalisation and technological development have increased the risk of narrow training for specific jobs, rather than broad cognitive, technical and professional capabilities that can be applied in a wide range of situations, and degree level qualifications that are recognisable worldwide. This is blurring the boundaries between traditionally separate sectors, institutions and modes of provision associated with academic and technical / vocational education, both in countries with decades of industrial decline and service growth such as England, and those with sustained industrial strength such as Germany.

For people with experiences of education and work rooted in binary systems, this may be characterised as academic or vocational drift, and their resistance may be compounded by the entrenchment of separate approaches to funding and regulation. There is, though, a long pattern of educational expansion through the integration of vocational and professional training with universities and higher education. I expect this historical evolution to continue during the next 30 years, with governments positioning technical education within unified tertiary education systems that promote coherence and progression between different types of learning throughout life.

(Professor Christopher Millward is Professor of Practice in Education Policy within the School of Education, University of Birmingham and former Director for Fair Access and Participation at the Office for Students)

Potential Future Scenario

In considering the insights outlined in the preceding chapters and the views provided by the diverse stakeholders, key themes and a view of the future of technical education for the next 30 years can be hypothesised. These also support the argument being made in the earlier chapters that the future of technical education will be a critical part of the wider future education system. The six key themes are outlined below:

Integration: There is an emphasisis on the need for greater integration of technical education into the broader educational landscape, removing unnecessary silos and evolving towards a system where technical and academic education seamlessly coexist. This integration would support the development of diverse and skilled workforce equipped to meet the needs of industries being changed by advances in technology.

Insight: There is a well-defined need for technical education to be agile and align with the needs of a changing labour market. This will require the ability to anticipate future talent needs, preventing over/under supply in key areas, and identifying skills gaps in the workforce. Employers have been identified as being central to preparing the workforce for inevitable ongoing change, with education and training providers needing to be responsive to their needs.

Inclusivity: There is a general commitment to inclusivity, with a desire to ensure everyone benefits from the opportunities provided by upskilling and reskilling. Lifelong and life-wide learning are highlighted as important concepts, ensuring individuals can access technical education at different points in their careers and within different settings/environments.

Vision: The need for a long-term vision for technical education is identified, with a desire to end a pattern of short-term approaches to policy. There is an aspiration for a shared vision and greater collaboration between industry and education providers. This is seen as a critical factor in successfully meeting future skills needs and providing innovation in the delivery.

Technology: There is a recognition of the significant and positive impact technology will have on the workforce and more generally industry and society. Technical education will need to evolve to meet the opportunities that will arise from future innovations. This will include understanding the key skills needed by the workforce to benefit from the technological advancements.

Sustainability: The growth of green industries and the wider sustainable economy is anticipated as key area for development within technical education. Investment by industry and education providers in new qualifications and learning opportunities is identified as being critical in achieving the desired outcomes.

The dominant view is for a technical education policy that is integral to a long-term economic plan, responsive to the needs of employers, inclusive, technologically agile and conscious of the wider needs of society. This vision is a policy environment where technical education plays a pivotal role in delivering a highly skilled and adaptive workforce, delivering economic growth and sustainability.

The preceding chapters and the stakeholder views identify a range of barriers to successfully realising a preferred policy landscape. These include:

- Market failure in predicting future skills needs, exasperated by the inability for effective collaboration between industry and education providers
- The lack of policy integration and policy initiatives that perpetuate a two-tier system between technical education and what is perceived as academic education
- An overemphasis on young learners
- The underutilisation of practitioners and learners in shaping the policy environment
- An overemphasis on competition between providers
- The need for political interventions and the impact of geopolitical issues

These barriers must be factored into the development of any future policy scenarios. Whilst some of these can be addressed through system leadership of the national system, there remains a need to adapt to the challenges posed by wider external influences.

The preferred policy scenario for technical education in 30 years is explored below, which takes into consideration the historical context, expert insights and the identified barriers.

There will need to be an integrated policy framework for tertiary education which ensures technical education fits seamlessly into the wider system.

Key policy levers:

- Collaboration between traditionally academic and technical providers is required and supported through quality and funding criteria.
- The framework will enable flexible pathways for learners, supporting a structured transitions between educational tracts.
- A common set of core skills and attributes identified across all disciplines and qualifications which are supported by the relevant regulators.
- Devolution of funding will support collaboration within regions and between types of education provider.

There will be co-creation with industry and a national approach to forecasting skills needs.

Key policy levers:

- Future job requirements will be supported by a robust national system for forecasting skills needs based on high-quality industry data.
- There will be regional industry advisory boards to co-create curriculum and support development.
- Partnerships between education providers and industry will be rewarded through funding initiatives and support for the expansion of provision.
- Professionals from industry will be incentivised to have continuous engagement with education and training.

An ethos of inclusive lifelong learning will be a distinguishing feature of the national industrial policy.

Key policy levers:

- The system is designed to enable flexible pathways and modes of delivery to support individuals to return to learning, including professionals and those in lower-skilled roles.
- Employers are supported to identify and target non-traditional learners to upskill or reskill early in the cycle of change within an industry.
- There will be financial incentives to support individuals to engage with continuous learning outside of the workplace.
- There will be structured relationships between local government, education providers, employees and employers to facilitate ongoing collaboration and innovation.

These strategies seek to shape a future environment where technical education is flexible, responsive, and available to a diverse range of individuals. This will facilitate the development of a skilled and agile workforce.

Conclusion

The evidence presented throughout this book and highlighted in this chapter is that the current trajectory for technical education lacks a well-defined roadmap or system leadership, supported by innovative system policies and sustained longitudinal interventions. It is also evident that there is a risk of perpetuating a cycle where decisions about the future are made for practitioners and the beneficiaries by individuals removed from the immediate skills challenges, sometimes possessing limited understanding of industry or the public service needs. Alternatively, these groups must demonstrate their ability to shape the future. It has been argued that the voice of the learner and employee needs to inform future policies, to better understand the skills required to meet the needs of evolving economy within England. If the preferred scenarios are adopted, in 30 years the country could potentially have a robust and responsive technical education ecosystem that is able to prevent the traditional barriers to innovation.

Epilogue

In this book, the authors have discussed how technical education is often perceived in England as inferior to academic education. Yet, successive governments, and most employers recognise how important it is to the economy and the well-being of the nation.

They have shown how technical education has developed over the last 200 years and in the wake of the Industrial Revolution and the central role that the state has taken in this process not least through numerous policy initiatives. Amongst the most recent of these have been the Skills and Post 16 Education Act, 2022 along with the introduction of T levels and the Lifelong Learning Loan. The authors have also shown that many of the challenges that face technical education today have their roots in the early industrial age.

The key challenges they believe that have faced technical education and continue to do so are:

- Policy change and churn with a lack of consistency
- Perceptions of the inferiority of technical education
- Neglect and underfunding of key providers—the further education colleges
- Lack of clear routeways for technical qualifications
- Inequalities in access both geographically and socially

In response, the remedies they have prescribed are:

- The key role of higher-level apprenticeships and their potential for expansion
- The role that further education colleges could play and how this could be achieved via supporting collaboration and more realistic long-term funding
- Clearer pathways for those taking techncial qualifications
- More effective collaboration between government and employers to identify the technical qualifactions needed now

The authors have suggested that as a result of our research, that FE colleges have a much greater role to play in the development of technical education.

They have discussed in detail the importance of apprenticeships as part of technical training and how their take-up has changed recently.

They have particularly emphasised the barriers that apprentices face if they wish to progress on to higher education and they have given our response as to how that can be remedied.

Similarly, the authors have described the difficulties that people have in accessing technical education at universities or colleges if they come from disadvantaged backgrounds, have disabilities, are from ethnic minority groups or are returning to education as mature students.

Against this backdrop they have suggested how the government along with employers can provide the infrastructure and the training to ensure equitable access to higher levels of technical education and a rejuvenated British economy.

References

Government of the United Kingdom. 2017. *Futures Toolkit Edition 1*. Available at: https://assets.publishing.service.gov.uk/media/5a821fdee5274a2e8ab579ef/futures-toolkit-edition-1.pdf (Accessed 14 January 2024).

Inayatullah, S. 2008. Six Pillars: Futures Thinking for Transforming. *Foresight* 10: 4–21.

Sardar, Z. 2010. The Namesake: Futures; Futures Studies; Futurology; Futuristic; Foresight—What's in a Name? *Futures* 42: 177–184.

Glossary

AOC—Association of Colleges The main body which represents further education colleges in England.

APP—Access and Participation Plans A document which must be produced annually by higher education institutions in England which wish to charge tuition fees of more than a figure prescribed (currently, £9250). The plans should demonstrate how the institution is going to spend part of the fees on supporting students from under-represented backgrounds.

Apprenticeships Apprenticeships combine practical training in a job with study.

Apprenticeship Levy All employers paying a wage bill of more than £3 million per year are required to pay 0.5% of their payroll each month as a levy tax. This levy can then be reinvested back into their workforce in the form of apprenticeship training.

BTEC—The Business and Technology Education Council An organisation that validates and assesses a wide range of vocational qualifications in England.

CHE—College Higher Education The standard term used to define higher education delivered in further education colleges in England.

J. Baldwin et al. (eds.), *The Future of Technical Education*,
https://doi.org/10.1007/978-3-031-70756-8

FAVE—Further Adult and Vocational Education An overarching expression for the education sector that delivers education beyond secondary school level.

Foundation Degree A combined vocational and academic qualification equivalent to two-thirds of a bachelor's degree which can be taken on a full- or part-time basis.

HND/HNC A combined vocational and academic qualification equivalent to two-thirds of a bachelor's degree. The HND is usually taken over two years on a full-time basis and the HNC on a part time basis.

Levelling Up Projects for local people and local businesses across the country to help grow the economy, create jobs, improve transport and provide skills training and local business support.

Lifelong Learning Entitlement A single funding system to help people pay for college or university courses, and train, retrain and upskill flexibly over their working lives.

Ofsted—The Office for Standards in Education An organisation responsible for inspecting educational institutions including schools and colleges.

OfS—Office for Students A body which acts as the regulator and competition authority for the higher education sector in England.

POLAR A system operated by the OfS which classifies local areas across the UK, according to the participation rate in higher education by the occupants of that area.

Polytechnics Tertiary education institutions offering higher education qualifications, which became universities in 1993.

Social Mobility Social mobility is linked to equality of opportunity: the extent to which people have the same chances to do well in life regardless of the socio-economic background of their parents, their gender, age, sexual orientation, race, ethnicity, birthplace or other circumstances beyond their control.

STEM The term used to group together science, technology, engineering and mathematics.

T Levels These are new two-year vocational courses which follow GCSEs and are equivalent to 3 A levels.

TEF—The Teaching Excellence and Student Outcomes Framework An exercise which is designed to assess excellence in higher education institutions.

Index

J. Baldwin et al. (eds.), *The Future of Technical Education*,
https://doi.org/10.1007/978-3-031-70756-8

SPRINGER NATURE

GPSR Compliance

The European Union's (EU) General Product Safety Regulation (GPSR) is a set of rules that requires consumer products to be safe and our obligations to ensure this.

If you have any concerns about our products, you can contact us on ProductSafety@springernature.com

In case Publisher is established outside the EU, the EU authorized representative is:

Springer Nature Customer Service Center GmbH
Europaplatz 3
69115 Heidelberg, Germany

The manufacturer's authorised representative in the EU is Springer Nature Customer Service Centre GmbH, Europaplatz 3, 69115 Heidelberg, Germany. If you have any concerns regarding our products, please contact ProductSafety@springernature.com

Printed and bound by CPI Group (UK) Ltd, Croydon, CR0 4YY
01/10/2026
02225898-0001